The Useful T

Northern N.g

GW01158852

Hugh Vandervaes Lely

Alpha Editions

This edition published in 2024

ISBN : 9789362092557

Design and Setting By
Alpha Editions
www.alphaedis.com
Email - info@alphaedis.com

Contents

PREFACE.

The text and plates of this volume have been prepared with the primary object of identification of the trees of the savannah forests of Northern Nigeria. The volume is, in no sense, a Flora, and no attempt has been made to include all the species found in the region under survey. Trees, being distributed over a large number of Families of the plant kingdom, are a small proportion of the total number of flowering plants, and their systematic arrangement would not, except in a few cases, be an aid to identification. Some 30 Families are represented by the 120 species included, in one of which there are as many as 36 species. The botanical names are arranged, therefore, in their alphabetical order so as to group certain genera and species whose similarity aids identification and to make the volume, in its index form, handier for reference.

The text is arranged in paragraphs, the first giving a general description of the form, height, girth, habitat, locality, &c., successive paragraphs describing in detail the bark and slash, thorns, wood, leaves, flowers, fruits and seeds, with a few notes on uses where these are known. These paragraphs are in uniform order for rapid reference.

The text is supplemented by drawings of the flower, leaf, thorn, fruit and seed, sketched from the living specimens. Wherever possible the drawings are actual size to avoid calculations from enlarged or reduced figures, but a number have been enlarged to show structures, or reduced owing to their actual size. In several cases sections of various parts have been made to illustrate peculiarities of structure, a pocket lens being used for this purpose, though practically all the parts drawn are visible to the unaided eye.

The majority of the botanical names have been verified through the Herbarium of the Royal Botanic Gardens at Kew.

In describing the woods, few of which are in use and many of which are little likely to be of use, an attempt has been made to strike a path midway between the complete technical and botanical description employed with regard to timbers and a vague statement as to colour, quality, hardness and grain which helps little to distinguish the individual from others like it. Most of the woods are described from samples collected by the author and shaped by the Sokoto Arts and Crafts School. I am indebted for these, and for the facilities offered for observing their behaviour under tools, to Mr. W.E. Nicholson. The dry weight of seasoned wood is given in all cases and in this connection it may be noted that a number of species produce heavier wood in the north than they do in the south.

A table of flowering seasons has been appended, with a few notes on its significance.

A second appendix arranges the 120 species under their Families. The *Genera Plantarum* of Bentham and Hooker has been followed.

Finally, there is an index of *Hausa* names for reference to the plates and text, whose numbers are given in the index. A blank column has been left in the index for the benefit of those who may desire to collect either additional *Hausa* names or the nomenclature of other tribes. It is hoped that the scope of the volume will be enlarged thereby, so as to cover other parts of Nigeria where the trees are the same but the language different.

The botanical name, being universal, is always given preference. The native names, variable and unreliable as they often are, should be a secondary consideration, their importance resting in their being a means of communication with the native rather than a short cut to the botanical names. In a country where there are several native names for one tree or one native name for several distinct species, the knowledge gained is either limited to a small number of trees or applicable to a small area of country. But once examined and properly identified, no number of native names need confuse the real identity. In collecting native names a warning should be given against the ignorant or ready-to-please native, and independent corroboration is advisable before accepting a name as worth recording. At the same time it is a fact that names genuinely differ in quite small areas of country where the same language is spoken, and a knowledge of the botanical names or real familiarity with the tree species is essential to reliable work.

The identification of 120 species, though seemingly not a large number, is sufficient to afford a thorough familiarity with most of the savannah forests and makes further species, not included here, stand out all the more clearly from their no longer similar companions. Numbers or economic values of further species are sufficient justification for forwarding these for identification to Kew, if they cannot be named from their resemblance to known species or by means of a Flora.

In selecting the 120 species to be included in this volume, the author has had to exercise his own discretion. The collection has been made between 9° and 14° N., and those who know this area will find a great deal they require, while some, in the more northerly latitudes, will find little of importance that is not included. Others, however, will meet with unfortunate omissions, for the following reason: South of 11° N. certain species which more truly belong to the evergreen and intermediate forest zones have established themselves in considerable numbers by streams, in valleys and in pockets of rich soil with surroundings suited to their propagation. Either all or none must be included if this volume is to be representative of such regions, but since they are not truly savannah species their inclusion here is beyond the scope of this volume, which would have to be enlarged very considerably to cover even the forests in the Benue region and south of the Niger, where savannah abounds. All the species included here occur up to 11° N., and with five exceptions up to 12° N., while nearly all occur below 10° N. and many much further south.

A selection has been made in the case of some genera, *e.g.*, *Ficus*, and *Combretum*, to familiarise the characters of trees which occur in large numbers and are typical of savannah forest. Some of these are of little importance beyond their occurrence in numbers over large stretches of country. The value of trees varies largely according to locality. Timber value is of little account in regions where it is inaccessible and unexploitable. Food, medicinal and other values take precedence where there is a large population demanding economic produce, and many species must be accounted valuable merely as a soil covering for the prevention of denudation, fixation of shifting surfaces, protection of new growth, grazing areas, precipitation of moisture, retention of conditions advantageous to agriculture or as a basis for the introduction of new, or the encouragement of existing valuable trees from the forestry standpoint. The advance of agriculture or the condensing of population intensifies these values by the destruction of the better types in favour of the poorer, and by the increased demand for the products of the better types. There are many instances of valuable products being obtained from forests which are not only some distance from habitations but whose soil is of a nature that precludes any possibility of supporting a population, since it will not produce crops. For example, *Sclerocarya birrea* (Danya) grows to large sizes in some parts of Sokoto Province where the soil is of an unworkable kind and where, over wide areas, there are no people. Yet this area is visited from all over the country and hundreds of trees are felled for the manufacture of mortars, trees in many cases suitable for two or three mortars being exploited for one only.

Though much has been written about the various types of savannah forests, a short account of them will not be superfluous here. First it may be pointed out that savannah forests vary so much in their composition and distribution that a survey is practically impossible, and if made, would have to be revised annually. Savannah is a particularly aggressive form of forest growth. Given the opportunity, it at once occupies more land and on its own ground the inferior types oust the superior wherever there is an opening. Any extensive area of cultivation, if abandoned and allowed to regenerate its tree growth, lets in secondary forests containing species which are less exacting as to soil and moisture conditions, and the depreciation of these conditions to the farming methods stops the competitive growth of the trees which formerly occupied the area. Where the area cleared is sufficiently small and the period short enough, the conditions are retained, and are able to close up the clearing with little loss by degeneration.

The fringing forests, belts of evergreen and deciduous trees on the banks of streams, are continually hard pressed by the broad wedges of open savannah between them. Fires take toll of them every year and cultivation not only pushes them down stream from the source but interrupts and cuts them up into islands exposed to threats of extinction.

Savannah is very topographical, and though, within broad limits, it can be divided into two main types, Tree and Bush savannah, the two are so distributed or blended that the differences are often unrecognisable and not able to be recorded in a survey.

Tree savannah is recognised as a tall type having a light, closed canopy, with a sprinkling of under-shrubs and a sparse growth of grass.

Bush savannah is a more open growth of less height, more spreading and lower branched form, with no canopy, and many stunted shrubs and a dense growth of grasses which are normally burnt out each year. There is every variation of this latter type from *Combretum* and *Guiera* scrub of barren soils or stony plateaux to the type which closely approaches that of Tree savannah.

The following are some of the recognised sub-types:—

PARK SAVANNAH.—This is a natural or artificial formation, the latter a product of agriculture, and is a two-storeyed growth composed of large type trees scattered about and dominating an undergrowth of smaller trees. Type species are *Parkia filicoidea, Butyrospermum Parkii, Afzelia africana, Tamarindus indica, Acacia albida*, etc. In the artificial product the large trees often owe their presence to their value, having been allowed to stand when the land was cleared for farms, the lower storey being a subsequent growth whose height depends on the period that has elapsed since the farms were abandoned. Natural Park Savannah is well represented by the superior size of *Afzelia africana* or *Paradaniellia Oliveri* over the lower storey of *Terminalia* and other species which cannot attain the height or proportions of these large trees.

FRINGING FORESTS.—These are the narrow belts of forest along stream banks, and are either intrusion of evergreen and other trees from a lower latitude or the remnants of broader belts which have dwindled to a mere fringe. Typical species are *Khaya senegalensis, Eugenia guineensis, Adina microcephala, Diospyros mespiliformis*, etc., with a number of evergreen shrubs, typically Jasmines. The narrower the belt the lower branched are the trees.

"KURIMI."—This is the formation found in depressions, where there may or may not be a stream, not necessarily flowing in the dry season. It is an enlarged edition of the fringing belt and more nearly represents the evergreen forest. Trees of large dimensions with long, clean boles are found amidst a luxuriant undergrowth of shrubs and younger trees, which if undisturbed will attain full maturity. "Kurimi" is the chief source of timber in the Northern Provinces and from it such type species as *Khaya grandifolia*, *Chlorophora excelsa*, *Albizzia Brownei*, *A. fastigiata*, *Diospyros mespiliformis*, and frequently Oil Palms are obtained.

"FADAMMA."—This is represented by the broad river valley perennially inundated, or a depression of varying size in flat country where water lies during the rains. Typical species are *Mitragyne africana*, *Paradaniellia Oliveri* and *Borassus flabellifer*, the Fan Palm. There is either a dense growth of grass or, in the case of temporary lakes, an area of cracked mud.

Pure forests are rare in savannah and *Acacia Seyal* and *Isoberlinia* species are type cases. Gregarious clumps of various sizes are, however, one of the commonest features and a large number of trees is concerned. Examples are the *Acacias Seyal*, *arabica*, *albida*, *Senegal* and *campylacantha*; *Anogeissus*, *Isoberlinia*, *Bauhinia*, *Anona*, *Bombax*, *Terminalia*, *Combretum*, *Detarium*, *Gardenia*, *Parinarium*, *Pseudocedrela*, *Stereospermum*, *Ximenia*, *Uapaca*, *Boswellia*, *Monotes*, etc.

Savannah also shows many examples of a dominant species, the presence of which are indications of the composition of a forest. The presence of this or that species sets a standard of comparative quality to an area by which it may be valued from various standpoints.

Over wide areas of savannah there are to be found many evidences of previous peoples in the shape of walled ruins, foundations of corn stores, grindstones and heaps of hand-picked stones indicative of cultivation. The dating of such remains gives the approximate age of the forest growth, and from this evidence it would appear that certain types of savannah deteriorate considerably with age and that farming will regenerate them to a certain extent. A comparison of 20-30 year old forest with the untouched older forest shows the former to be denser, more healthy and of straighter and more even growth, while the latter contains a large number of very old trees in a state of decay, with great open crowns, liable to be blown down by storms, their branches burned by fires and riddled with fungus. The new growth which is to take their place consists mostly of grasses, stunted shrubs and small trees of species which are incapable of forming a canopy. As mentioned above, provided the area farmed is not too large, the support given it by the surrounding forests is sufficient to ensure effective regeneration, otherwise a poorer type takes hold.

In conclusion a short account of the growth and characteristics of savannah forests is given. As would be expected, the trees, being subject throughout life to extremely hard conditions, counter with defensive measures for self-preservation. Most of them are prolific seeders and many have the habit of retaining their fruits or seeds on the tree for many months, sometimes right round to the next flowering season. Germination of seeds is a matter of chance when the tree is liable to ground fires from October to April or May. The life of a seedling, too, is precarious, since at the age of six months it may be subject to a devastating fire. In later life a fire, occurring in April, may, if the grass is high, burn the bark to ashes on the outside, destroying all fruits or seeds, killing the twigs and small branches and apparently destroying all life. Yet, a month later leaves will spring from all but the burnt tips, the tree losing a year's height growth, but in a position to put on another year's stem girth. Wounds on the trunk of a tree heal over with the formation of a hollow or rotten core. Trees 4-5 feet in girth have been felled and their stems found to consist of a mere shell two or three inches thick. The hollow is filled either with the workings of termites or the fermenting sap which is forced up from the ground level and will pour out of the stump. Savannah trees are adapted to overcome most of these adverse conditions. Heavy crops of fruit and seed, rapid ripening of seeds or fire-resisting fruit coats and various means of distribution are means to ensure germination. Deep or thickened tap roots of seedlings are put down in the first year so that though the seedling may be levelled by fire it will appear with renewed vigour soon after. A very short flowering period for the individual tree is a marked characteristic of many species. Fire has a quickening effect on the flowering and leaf bearing of trees to a marked degree. A grass fire in November may produce flowers and leaves in December on trees which, if unburnt, would not have flowered till February. As a rule, no tree will flower again if the first flowers have been burnt off, though it will, of course, bear leaves. One or two exceptions, very rare, have been noticed by the author, but these were due to the peculiarity that some trees exhibit of bearing flowers on some of their branches and not on others, so that the flowerless twigs were probably excited by the fire into bearing flowers after it. Leaves are readily replaced after attacks by locusts which generally take place early in the season. The end of the cold weather and first sign of heat is the spring of growth. Few trees have to wait for rain before they produce all the signs of maturity except the complete growth of their leaves. The bark of young seedlings, subject to fire, is very thick and corky and that of old trees is still more so, while the inner layers are fibrous and full of sap. Almost all species will coppice very well, many throwing up shoots up to 10 feet high in a season, others only a foot or two. Root shoots are also very common and a large number of what are apparently seedlings in the forest are root shoots. Most trees are anchored very firmly to the ground by lateral roots far longer than the height of the tree, a protection against storms. A tree which is completely ringed will often endeavour to join up the cut, meanwhile continuing its season's foliage, while the smallest connection with the root is sufficient for it to live for many years. Most new leaves, as a protection against the sun, are reddish in colour, many a brilliant crimson, the green being produced through all gradations of colour from the red. Others are hairy, scaly or covered with a bloom, the protective coverings wearing off with maturity. The proportion of sapwood to hard is large and many species show marked differences in this proportion according to their locality, some showing no heartwood in the north where it is produced in the south. The weights of woods are often considerably heavier in the north than is the case for the same species in the south. There are some localities where wide areas are infested with termites to such an extent that, during the dry weather their whole surface, even to the tips of the twigs, is completely covered with the earth carried by the termites. Although there are exceptions, it is observed that apparent damage is negligible and that the old scales are removed from the bark, exposing the fresh surface.

The savannah species are, then, well equipped to withstand fire, drought, insects, wounds, storms, and damage by man and it can be imagined how readily they will supplant better types, since the harder the conditions the better adapted are the trees to meet them. The wide distribution of most species renders them more aggressive, and an individual species that will grow into a fine tree under the best conditions can still grow under the worst, though it may differ in form and feature so as to be hardly recognisable as the same species.

If, as it is supposed, savannah forests are depreciating of their own accord, some factor other than climate, which has not been demonstrated by records, must be at work. It is sufficient to point to the increase and spread of the population from the towns to the forests to find a reason for the displacement of good forest by lower grades. Shifting cultivation encourages a worse type each time it is practiced and late fires are very destructive.

For valuable assistance given in identifying the trees and shrubs in this volume, the author is indebted to Dr. A. W. Hill, Director of the Royal Botanic Gardens, Kew, and Mr. J. Hutchinson of the Herbarium. To Dr. J. M. Dalziel he is indebted for the "Hausa Botanical Vocabulary," a valuable short-cut to further research. Other books referred to are the "Flora of Tropical Africa" and Kew Bulletin, Additional Series IX., neither of which has been used other than for purposes of classification, it being the author's aim to describe and illustrate the material as it appears to be, and not only as it actually is from the scientific point of view, avoiding botanical terms except where they have no alternatives.

H. V. L.

March, 1925.

ACACIA ALBIDA Delile.—*Gawo*. LEGUMINOSAE.

This is the largest of the Acacias in the north, and attains a height of over 60 feet with girths over 12 feet. It is particularly common in Sokoto, but even there is locally distributed, and gregarious clumps are commonly met with. It prefers dry, sandy soils. The form of the young and the old tree is very different. The former has a straight bole with acutely ascending branches forming a high, flat-topped crown very similar to that of *Paradaniellia* in shape. The bole may be clean for over 20 feet, but in young trees there are usually clusters of thorny twigs at no great height. As age increases the crown widens and the limbs get heavier and more spreading until the form is like that of *Parkia*. The bole is then quite clean and considerably thickened at the base, the roots spreading above ground and forming flanges, thick but of little height. Its distinguishing peculiarity is the habit of shedding its leaves at the approach of the rains and putting them forth at the first sign of the dry season, in September. It grows fast up to full height and then slowly.

The Bark is uniform dull grey. Long wide fissures and prominent ridges of hard bark ascend the bole and the bark appears as if it was stretched without being cast off. Light brown patches, quickly turning grey, are left by the falling scales. The slash is pale brown and fibrous.

The Thorns are in pairs, under ½ inch long and slightly recurved. They are pale brown with white bases. The young tree is armed all over, but as the tree ages the thorns leave the stem and finally the higher branches are free as if protection seemed to be unnecessary at a height.

The Wood.—The heartwood is a light, clean yellow. The sapwood is a dirty white. Frequently the whole wood is a dirty white or grey, but this is due to discolourations from mould and the wood should always be seasoned in dry air and not allowed to get wet. In transverse section the rings are indistinct and wide apart, the pores are small and evenly distributed, mostly in festoons with twin pores here and there, the soft tissue plainly seen as concentric lines with the hard tissue alternating. The rays are extremely fine, very regular and closely spaced and quite invisible to the naked eye. In vertical section there are faint bands of colour, with light reflecting flecks in tangential section. The wood is soft, very easy to work with all tools, seasons well if looked after, and finishes well under the plane. The weight is 35 lbs. a cubic foot.

The Leaves are bipinnate, 4 inches long with 6-7 pairs of pinnae bearing some 10-15 pairs of dusty, grey-green leaflets, covered with tiny hairs. The foliage in the mass appears bluish-green. It appears at the end of September and is susceptible to insect, caterpillar and locust.

The Flowers, from October to December, are in dense spikes 3-4 inches long, scented. The lower inch of the spike is flowerless. Several spikes are borne at the twig ends. Each flower has a 5-lobed cream-coloured calyx with pink centre, 40-50 stamens, and a short pistil with slightly clubbed stigma.

The Fruits are pods, at first green and sickle shaped, ripening to a bright orange and twisting into strange shapes. They are 4-6 inches long, 1 inch wide, concave on one side, convex on the other. They contain dark brown shiny seeds ⅜ inch long. The pods ripen in January and February and fall entire, rotting on the ground.

Uses.—Inferior canoes, of the stitched together type are made. The ripe pods are collected and fed to cattle, sheep and goats. The common brown kite nests frequently in its branches in March and April.

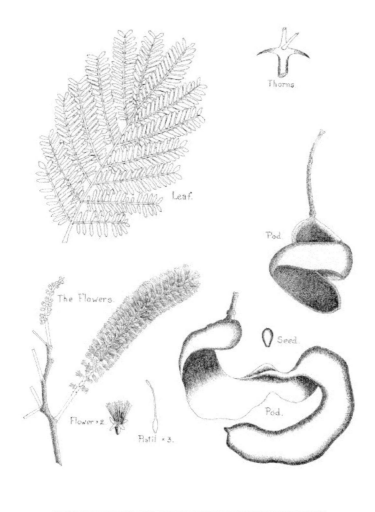

ACACIA ARABICA Willd.—*Gabaruwa, Bagaruwa.* LEGUMINOSAE.

This tree, the original source of gum-arabic, is very common and widely distributed throughout the north. It is very frequent round towns where it is utilised for the tannin properties of its pods. It varies largely in size from a small erect tree with umbrella crown to a large, heavy-stemmed tree with high rounded crown. It occurs gregariously in clumps or small forests, growing densely with the crowns meeting overhead. The seed germinates freely in low-lying country liable to submergence by rains or floods and the tree flourishes in such situations. It is just as partial to dry sandy soils where it abounds. With its almost black stem and branches, bluish-green foliage and graceful, symmetrical form, it is readily distinguished from other *Acacia* species. The average height is some 20-25 feet, but trees over 35 feet with girths up to 8 feet are not uncommon.

The Bark is almost black, or a dull dark grey, with long stringy ridges and narrow fissures. The dark colour extends to the branches and woody twigs, the latter covered with a soft pubescence. A clear yellow gum exudes from the slash which is a pale pink colour. The slash is red brown and blackish in streaks.

The Thorns are in pairs on the branches and twigs, some quite short and slightly curved, others, especially on the older wood being long, straight, slender and very sharp, with a grey colour and a slight backward slope. They are up to three inches in length and generally curved a little.

The Wood is a deep red-brown with almost purple bands. The sapwood is yellow. In transverse section the colour is darkest, the rings show as dark, ill-defined bands, the pores are small, few and scattered about between the fine straight rays which are visible against the dark ground of the hard tissue. There are many double and nested pores and zones with very few pores. Their contents nearly close them. In vertical section the grain is twisted, there are dark bands and the pores have black contents. The wood is very hard to saw and plane but the finish is hard and takes a high polish. The grain picks up badly in places. The transverse section will polish. The weight is 75-80 lbs. a cubic foot.

The Leaves are bipinnate, some 2-4 inches long with 5-6 or sometimes more pinnae bearing 15-20 pairs of leaflets with rounded tips, $\frac{3}{16}$ inch long. They are a dark bluish-green with grey bloom and soft texture.

The Fruits are jointed pods, straight or sickle-shaped, with slightly embossed seeds. The pod surface is wrinkled and covered densely with a whitish bloom. The seeds, some 10-12 in a pod, are round, flattened, shiny and brown, $\frac{3}{8}$ inch in diameter. The pods, ripening towards the end of the year, fall to the ground entire. They are the "Sant Pods" of commerce.

Uses.—Hoe and axe handles are made from the wood which rarely reaches large enough dimensions to provide canoes as it does in other parts of Africa.

A concoction of the pods, crushed with water, is used for tanning. The leaves, boiled, with the addition of a small piece of Tamarind pod, are used as a cure for a disease of the eye which causes the lashes to fall out.

From the pods a black leather dye called "kuloko" is made.

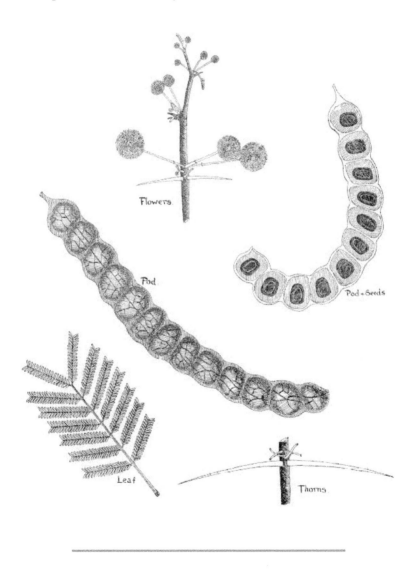

Flowers.

Pod.

Pod & Seeds

Leaf

Thorns.

ACACIA CAMPYLACANTHA Hochst.—*Kumbar Shafo, Farichin Shafo, Karo, Karki.* LEGUMINOSAE.

A tall tree somewhat resembling *Acacia Sieberiana*, especially in the case of full-grown trees of both species. A comparison of the bark, thorns and pods will readily distinguish one from the other. It is the tallest and most erect of the acacias in the north, reaching a height of 60 or more feet and girths of 4-6 feet when growing, as is very commonly the case, in dense clumps on stream banks, hillsides and on the sites of old towns. This last peculiar situation is sometimes explained locally by the fact that cattle eat the pods and deposit the undigested seeds on the site of a cattle camp in a deserted town. The bole is often 30 feet or more in length in these clumps, the crowns high and flat-topped and meeting overhead, forming a density of shade sufficient to kill all growth of grass on the floor. Old trees in the open have lower, wider-spreading and more open crowns, with shorter boles and bear a marked resemblance, at a distance, to *A. Sieberiana.* They commonly reach over 6 feet in girth.

The Bark is a pale yellowish colour, sometimes almost white, and is smooth, with small, regular, brownish scales which in old trees are grey and coarser. The slash is crimson with white streaks, very fibrous.

The Thorns, which are the readiest means of distinguishing this species from *A. Sieberiana*, are short, strong, recurved, and brown with a black point. They resemble falcons' claws, from which they get their native name. They are in pairs at a widely obtuse angle.

The Wood is a very dark brown, with almost black streaks. The sapwood is white. In transverse section the rings show as irregular dark bands, the pores are small, few and connected by very thin lines of soft tissue, the rays are very fine, waved and unevenly spaced, invisible to the naked eye. In vertical section the rings show as bands of dark brown and almost black and there are lighter areas. The wood is hard, fibrous and bad to saw, not easy to plane, though the finished surface is smooth and will take a good polish. The long fibres pick up badly in places. The weight is 52 lbs. a cubic foot.

The Leaves are bipinnate, about 9-10 inches long with 20-25 pairs of leaflets which droop on each side of the mid-rib. The narrow leaflets are $^3/_{16}$ inch long, slightly curved and a dull, dark green.

The Flowers are in 4-5 inch spikes about $^1/_2$ inch in diameter, densely crowded with cream-coloured scented flowers with numerous stamens. They appear from May-July in masses amongst the leaves.

The Fruits are pods, 4-5 inches long, $^3/_4$ inch wide, very flat, with the seeds slightly embossed. They are very numerous, tenacious and conspicuous on the trees from October onwards. The seeds, about six in a pod, are round, flat, shiny and dark brown, the same colour as the pods, and have strong germinating power.

Uses.—The wood is used for implement handles of all kinds. The branches are cut for making protective farm fences, and the pods are very much appreciated by cattle.

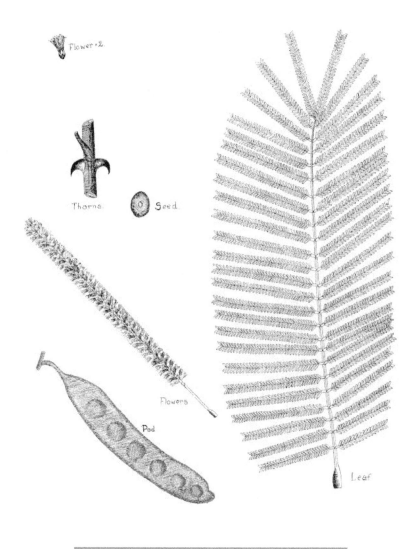

ACACIA DALZIELII Craib.—*Gaba chara, Gwanno.* LEGUMINOSAE.

This is a slender Acacia, locally very plentiful, especially in central and south-west Sokoto throughout some 1,000 square miles. It inhabits the open bush savannah of better quality where the rainfall is good and the soil contains some loam. In certain belts it is the predominant species. It is the largest leaved of all the Acacias in this zone and at a distance closely resembles *Entada sudanica*, though the leaflets are much smaller and more numerous and the flowers and thorns are both distinctive; in fact, the resemblance is one of form only. It grows 20-35 high with a girth up to 3 feet. Some trees are tall and slender with high crowns, others low-branched and spreading, but the foliage of all is very light and graceful.

The Bark is grey-brown with long fissures and large, shaggy scales which fall in long, irregular sections. This is largely due to fires which thicken and blacken the cork. The bark of the twigs is a silvery grey. The slash is a dull red, exposing the orange colour of the wood.

The Thorns are in pairs on the twigs, the leaves springing from between them. They vary considerably in size, 1/4-3/4 inch long, large and green on the new shoots and small and black on the wood, as a rule. They are slightly curved, and grooved along the inner side from the point where the leaf-stalk springs.

The Wood.—The heartwood is reddish with long, vertical streaks of black and brown in the pores. The sapwood is yellow with an orange grain. In transverse section the rings are indistinct, the pores are open and numerous, the soft tissue in wide and narrow festoons plainly visible to the unaided eye. The wood is hard, splits easily, is straight-grained, not easy to plane and weighs 65 lbs. a cubic foot.

The Leaves are 12-18 inches long, bipinnate with 20 or more pairs of pinnae bearing 50 or more pairs of long, narrow, pointed leaflets with parallel edges. Near the base on the dorsal side of the stalk is a prominent oval gland. They are a very bright, fresh green, and at first erect, finally spread and droop.

The Flowers are in large, erect panicles, 12-18 inches high and stiffly branched, standing prominently up on the ends of the shoots. They appear in the rains and are in bright yellow balls about 1/2 inch in diameter.

The Fruits are pods, 3-5 inches long, 1/2-3/4 inch broad, flat, dark red-brown with a grey bloom and containing from 6-12 flat, oval, brown seeds. The pod is slightly embossed at the seeds and very persistent, numbers remaining on the tree till the following rains.

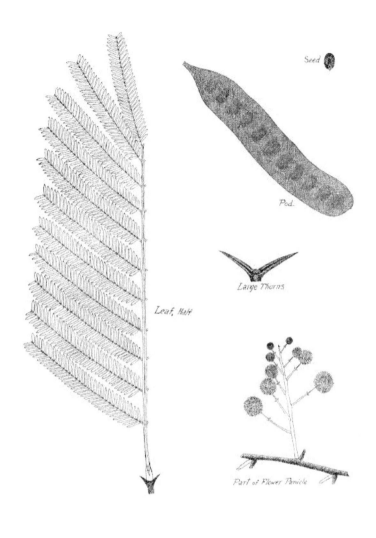

Seed

Pod.

Large Thorns

Leaf, Half

Part of Flower Panicle

ACACIA NILOTICA Del.—*Bagaruwa namiji.* LEGUMINOSAE.

This species is nearly allied to *Acacia arabica* and can be at once distinguished from it by the smooth, jointed pods. In other respects it is very similar. The native distinguishes the two species by the fruits, the *Acacia arabica* with its large grey "sant pods" being known as the female, "ta mata" as against the name "namiji" of *A. nilotica.* As a rule this species is a much larger tree than *A. arabica* and two forms are commonly met with. The one, which is common in low-lying country liable to inundation, has a short bole and a large number of slender branches which ascend to a great height and spread out wide, forming a large semi-spherical crown almost reaching the ground. The other has a long bole with a girth of 10 feet or more and a high rounded crown. The latter type yields a large volume of timber.

The Bark is almost black with deep fissures and very long, ragged scales, which fall in large pieces. The slash is red-brown, with darker streaks.

The Thorns are in pairs, quite small near the twig tips and up to 2 inches or more in length elsewhere, straight or curved, more often the latter, white, slender and very strong and sharp.

The Wood is reddish-brown in colour, almost blood-colour in cross section, with marked rings. It is very close grained, the cross section being able to be planed quite smooth. The grain is irregular and picks up. The weight is about 75 lbs. per cubic foot. See *A. arabica*.

The Leaves are 2-3 inches long with some five pairs of pinnae bearing 10-20 pairs of leaflets, bluish-green with a grey pubescence.

The Flowers are very numerous from February onwards in clusters of 2-5 at the nodes, the slender stalks over an inch long and the sweet-scented, yellow flower balls over ½ inch in diameter. As a rule the ends of the twigs on which the flowers are borne have no spines, these appearing later.

The Pods are very variable in size, up to 6 inches in length and their peculiarity is the marked jointing which, before they are full grown, gives them the appearance of a string of beads. They are smooth, with a slight bloom and contain some 6-8 oval, flattened seeds which ripen from March onwards.

Uses.—The wood is used for axe, hoe and tool handles.

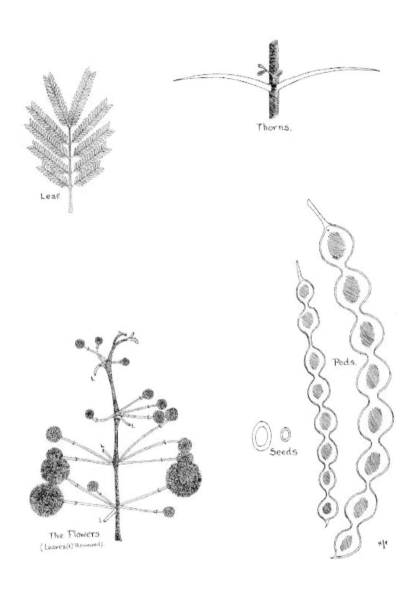

Thorns.

Leaf.

The Flowers
(Leaves(t) Removed).

Pods.

Seeds

ACACIA SENEGAL Willd.—*Dakwora.* LEGUMINOSAE.

A small tree some 15-20 feet high on an average, which is commonly met with in the more northerly provinces, especially in Sokoto, where it grows in dense thickets. It branches low down and is often shrub-like in its young stages, later producing a bole some 6 feet in length. The branches, ascending at an angle of about 60 degrees, repeatedly fork and form an open, flat-topped crown from which some of the long straight twigs protrude some feet above. It is a source of gum-arabic.

The Bark of young trees is very light in colour with a creamy tint, that of older trees, especially those which grow in the open, is purple on the bole, with patches of the lighter tint here and there. All over it are minute, whitish scales, so fine that they rub off in the hand as a fine powder, like that of *Acacia Seyal*. Here and there, according to season, are larger, thicker grey scales, chiefly about wounds or on the swollen forks. The bark is very thin and if scratched with the nail shows the bright green cambium layer just under the surface. The slash is mottled red.

The Thorns are in threes at the swollen nodes. The centre thorn is sharply recurved like a claw, the two side thorns being almost straight and pointing forward towards the tip of the twig. All are short, very sharp, with broad bases, like rose thorns, dark brown to almost black in colour, with a greyish bloom.

The Wood is white, but of such small dimensions that it is not used.

The Leaves are bipinnate, 3-4 inches long, with some six pairs of pinnae, each with some 20 pairs of leaflets, blue-green in colour and paler below than above. The ribs are covered with very short hairs.

The Flowers which appear in April are in 3-4 inch long spikes, one or two at the nodes. From ½-1 inch of the spike is flowerless, the rest densely covered with small creamy-white flowers, each having a pale green 5-lobed calyx, five pale green petals and a large mass of short, white stamens with a pistil indistinguishable amongst them.

The Fruits are pods, varying in size according as they contain 1-6 seeds. They ripen in November and remain on the tree till April, often a very heavy crop on the leafless tree. They are 1½-4 inches long and an inch broad, flat, slightly shiny, embossed at the seeds, sometimes the same width throughout, sometimes indented on the margin between one seed and another, sandy in colour, with often blotches of a darker colour. They split either whilst on the tree or fall entire and open on the ground, the seeds remaining attached to each half alternately and the two halves attached at the stalk. The seeds are round, flat, green-brown, with a U-shaped scar on each side and attached to the pod edge by a short, thick stalk. The pod is prominently cellular veined.

Uses.—The bark of the roots is twisted into ropes which are of great strength.

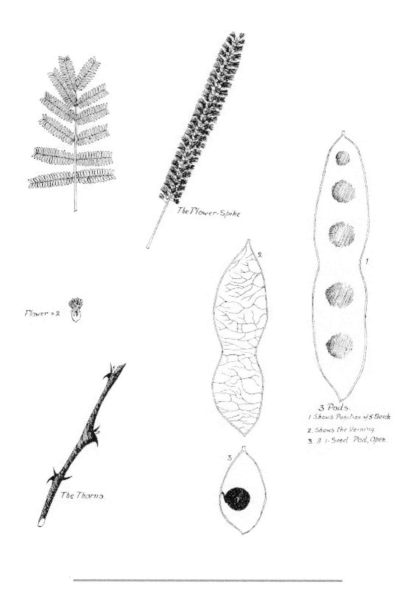

The Flower-Spike

Flower × 2

The Thorns

3 Pods.
1. Shows Position of 5 Seeds.
2. Shows the Veining.
3. A 1-Seed Pod, Open.

ACACIA SEYAL Del.—*Dushe, Dussa, Jimshi, Erafi, Gishishiya.* "*Talh*."
LEGUMINOSAE.

This very common species is noted for its occurrence as pure forest over
large areas of country. It can be distinguished at a glance from *A. Sieberiana*,
which it resembles in small trees, by its powdery orange or rust-coloured
bark, or in the case of the variety *A. fistula*, by the colour being a milky white.
It is generally a small tree from 15-30 feet in height, resembling *A. arabica* in
form, with an umbrella-shaped or flat-topped crown formed by the
ascending and spreading branches. Sometimes the secondary branches are

practically horizontal and the crown wide and quite flat on the top. It bears a profuse mass of bright yellow flower-balls, highly scented and very conspicuous. It occurs in clumps or isolated examples on stony ground, not in loose sand, but where the ground is broken up and barren looking.

The Bark, the most distinguishing feature, is powdery and comes off in the hand with the appearance of minute flakes like bran. It is a rust-red or orange colour and in the case of the variety *A. fistula* the bark is green with the powder milk-white in colour. A clear white or yellow gum exudes from the slash, which is light red and white, with brown edges. *A. fistula* is light red with green edges.

The Thorns are similar to those of *A. arabica*, but shorter, 1-2 inches long, white with black points, straight, strong and sharp and at an angle of about 100° to each other. The base is frequently largely swollen. Near the tips of the twigs little recurved thorns occur in place of the long, straight ones.

The Wood is whitish, a brown colour being given it by the lines of hard tissue. It is apt to discolour with mould. In transverse section the close concentric rings of hard and soft tissue are very distinct, the pores are very numerous in the bands of soft tissue of varying width, the alternating bands of hard tissue being almost free of pores. The rays are straight, the larger plainly visible, the finer often closer together, showing as light-reflecting bands in radial section. In tangential section the hard tissue makes a well-defined grain of light brown. The wood is soft, the grain coarse; it works easily with tools, the plane giving a rough finish. Weight 50 lbs. a cubic foot.

The Leaves are bipinnate, about 2 inches long with 6-8 pairs of pinnae cut up into some 15 pairs of leaflets. There are typically three in the angle of each pair of thorns and they are dark green when full grown.

The Flowers are in yellow balls about ½ inch in diameter, 1-4 in the axil of a leaf. They have 1½ inch stalks, are sweet scented and appear from January onwards in masses which make the tree all yellow at a distance. The young leaves appear just as the flowers are going.

The Fruits are slightly curved, flat, jointed, embossed pods, 3-5 inches long, light brown with green tinges and finely veined. When ripe the pod splits up both edges, and the seeds, 6-10 in number, remain fastened to the edges by the long twisted attachments, from which they break loose gradually, the pod sections remaining attached to the tree for some time before this occurs. Many pods fall with the seeds still attached to them. The clusters of curved, split pods are most conspicuous and they contract considerably during the ripening stages.

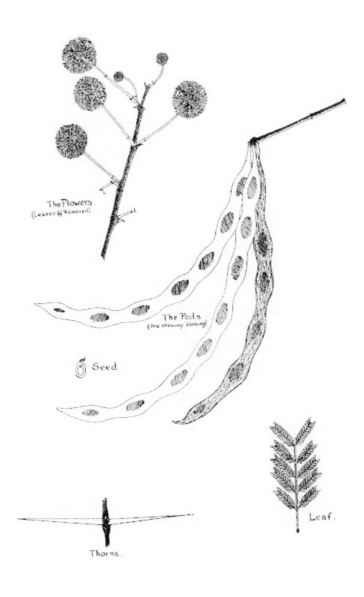

The Flowers.
(Leaves removed)

The Pods.
(One shewing Veining)

Seed

Thorns.

Leaf.

ACACIA SIEBERIANA DC.—*Fara kaya, Bauji.* LEGUMINOSAE.

This common and widely distributed species has, when full-grown, some similarity to *Acacia campylacantha*, especially when the species occur in the open. Its long, white thorns distinguish it from the latter. It is occasionally met with as pure forest, but should not be confused with *Acacia Seyal*, the "Talh" acacia, which is very like the small *A. Sieberiana* and has yellow flower-balls and ochrey bark. Old trees are a height of about 50 or more feet with girths of 5-6 feet. They have large round crowns and sometimes rather persistent lower twigs, though 20 feet boles are not uncommon. It grows well in dry situations and frequently occurs, mixed with *A. Seyal*, in open, dry country.

The Bark of young trees is yellowish and smooth, this feature persisting on the branches of older trees, which have rough, small, square scales, grey in colour on the stem and larger limbs. The bark exudes a gum which is white, clear and brittle when dry, making a fair mucilage. The slash is yellow, with dark red edges.

The Thorns, persistent everywhere except on the bole and largest limbs, are fine, straight and white, with very acute points. They are in pairs at about 120° to each other, pointing slightly forward, up to 3 inches long. On new shoots they are green and soft, hardening as the shoot ages. Quite small thorns are found at the tip of the shoots.

The Wood is a dull yellow colour, very subject to bluish discolorations, due to mould. In transverse section the rings are indistinct and inseparable from the numerous concentric lines of hard and soft tissue which are clearly marked. The pores are open, very numerous, of different sizes, the large ones plainly visible, mostly single with little soft tissue, except along the rings where the pores are in rows, each separated by a ray. There are a few double pores, nests and small chains. The rays are slightly waved, fairly evenly spaced except where the finer rays are closer together. The rays show as light-reflecting bands in the radial section. The wood is soft, rather coarse in grain, easily worked, planing to a rather rough, untidy surface. The weight is 45 lbs. a cubic foot.

The Leaves, which spring from between the thorns, are bipinnate, about 5-6 inches long with 10-25 pairs of pinnae, the numerous and dark green leaflets being ⅛ inch long with rounded tips.

The Flowers are in white balls slightly over ½ inch in diameter, with ¾ inch stalks, in loose bunches. Their colour distinguishes them from those of other species similar and likely to be confused. They are sweet-scented and visited by bees.

The Fruits are large, brown pods, from 6-7 inches long, 1 inch wide and ½ inch thick, straight or slightly curved. They are not very numerous, but conspicuous by their size. The seeds, about 12 in number are quadrangular, rounded, flattened, very hard and dark and light brown.

Uses.—The young trees and the branches of the old are used for handles of implements, the angle of branching being especially suited to the shape of the large hoe (galma) handle.

The gum, in solution, is applied to turban and gown cloth and a sheen produced by beating.

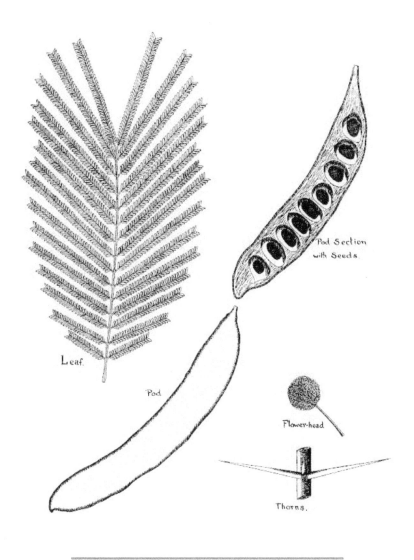

Leaf.

Pod.

Pod Section with Seeds.

Flower-head.

Thorns.

ADANSONIA DIGITATA B. Juss.—*Kuka.* "*Baobab.*" MALVACEAE.

The Baobab tree is so well known as hardly to need description. The enormous girth up to some 50 feet, large white flowers and pendulous fruits are familiar. From its earliest years it assumes the shape, in miniature, of the older trees. Though nature plays the largest part in the shaping, the cropping of leaves for food, the stripping of bark for rope and the ringing of large branches by beetles till they fall to the ground, all help to accelerate the abnormal form. Widely distributed, it is locally more than generally common, and abounds in groups. The majority of large native towns are full of it. Occasionally it can be seen without the usual stunted appearance, with slender branches and well-formed crown. An association with *Tamarindus indica* (Tsamiya) is fairly common, the latter growing long slender stems about the former and partly embracing its trunk.

The Bark is grey, with all shades of purple and a sheen. It has great callus-growing properties and appears vigorously alive in its power to cover up wounds. The outer bark is soft, spongy and full of sap, and it is the inner layers which are stripped off for the rope making. The slash is mottled red and white.

The Wood is not used. It is very light, soft and crumbling, rotting rapidly under exposure.

The Leaves are digitate with some six or seven lobes, 2-3 inches long, borne on a 4-5 inch leaf-stalk. They have sinuous margins and a tongued tip; the mid-rib is sunken and the veins regular. The surface is dark green above and smooth, but downy beneath. The leaves appear soon after the flowers.

The Flowers are solitary and pendulous on 9-10 inch stalks, and appear in May before the leaves. They are some 6-8 inches in diameter with five leathery sepals covered densely on the inside with straight hairs; five white petals nearly twice the length of the sepals, recurved at the tip and with wrinkled edges; a stout, shiny, white, tubular stamen column from which the mass of white stamens with light brown anthers radiate and bend up towards the vertical position and from which emerges the long shiny white pistil with spiral bend and outward growth, bearing at its tip the flattened, lobed, sticky stigma. All the flower parts tend to assume the vertical position. The Hausa calls them "kumbali."

The Fruits are large, oval or round, 5-15 inches long and 3-7 inches in diameter, covered with brittle, bronze hairs which break off when handled. The stalk is long and stout and the calyx remains at the base are broken and hardened. There is a short woody "nose." A number of kidney-shaped seeds are embedded in a white, crisp, acid and slightly refrigerant pulp, pleasant to taste when fresh, and fibres separate the rows of seeds. The seeds are grey with a brown patch, intricately veined. They are very hard. The pulp is called "garin kuka," the seeds "guntsu."

Uses.—The leaves are used as a sauce in soup and food, called "Miyan (or) Garin Kuka." They are given also, with bran and salt as a horse medicine, called "chusar doki."

The inner bark is stripped and twisted into strong ropes "kista," tethering ropes "gindi," and strings of musical instruments. The acid pulp is eaten fresh.

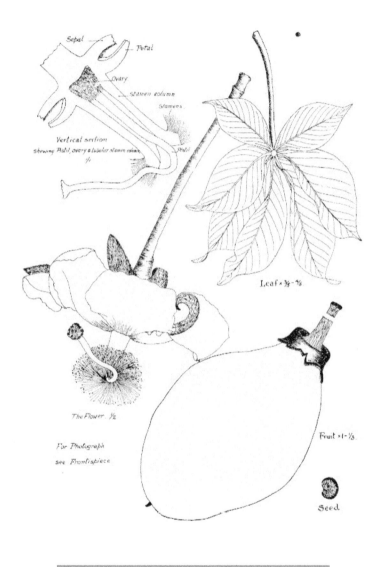

Sepal

Petal

Ovary

Stamen column

Stamens

Vertical section
shewing Pistil, ovary & tubular stamen column
½

Pistil

Leaf × ¼ – ⅛

The Flower ½

For Photograph
see Frontispiece

Fruit × 1 – ⅓

Seed

ADINA MICROCEPHALA Hiern.—*Kadanyar rafi*. RUBIACEAE.

This is a large tree inhabiting stream banks which extends as far north as 12°. It attains a height of 60 feet with a girth of 6-8 feet, occasionally more. It has a habit of growing right in the beds of small streams where it is washed annually by the flood waters. In these situations the trunk will assume almost a horizontal position at the base and survive severe damage. Except in dense stream bank vegetation the bole is rarely clean, but is covered with slender shoots down to the ground level. The erect willowy shoots are a marked feature of the tree. It has a large dense crown with heavy limbs, and the rosette-like growth of the long narrow pointed leaves distinguishes it.

The Bark is grey and deeply fissured, often spirally, with long fibrous ridges and ragged scales which fall in large pieces. The slash is a deep dull red, with thick spongy fibres, and a sticky sap exudes.

The Leaves are some 9 inches long and 1½ inches broad, with tapering base and long pointed tip. The stalk is an inch long. The upper surface is a dark, shining green, paler beneath with the mid-rib prominent. They are whorled and appear like rosettes.

The Flowers are borne on 3-inch stalks in the leaf axils from February to June. Each stalk has a pair of bracts in the middle, or nearer the head of flowers. These flower-heads are spherical, 1-1¼ inches in diameter, greenish and perfumed. Each separate flower is tubular, with a short 5-lobed calyx, a long tubular 5-lobed corolla, 5 stamens inserted on the corolla throat and a long, knobbed pistil. The corolla tube is pinkish, the tips of the lobes mauve, and the calyx is green, red inside at the base.

The Fruits are in spherical heads, greatly resembling those of *Mitragyne africana*. Each is a small capsule which parts down the middle into two cocci which contain a number of minute seeds, winged at both ends.

Uses.—The timber is suitable for all kinds of furniture and joinery work.

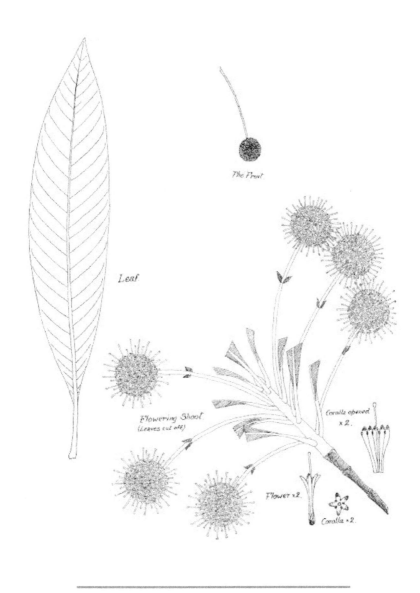

Leaf

The Fruit

Flowering Shoot
(Leaves cut off)

Corolla opened
× 2.

Flower ×2.

Corolla × 2.

AFRORMOSIA LAXIFLORA Harms.—*Makarfo, Kariye gatari.*
LEGUMINOSAE.

A small or medium-sized tree very common in bush or tree savannah, averaging 30 feet in height, but attaining 50 feet, with girths of 3-6 feet. The bole, though frequently of good length, is rarely straight, in fact it is a characteristic of this species that the bole is bent and twisted and that the branches show this feature to the tips. It is like a large edition of *Stereospermum*

Kunthianum (Sansame). The crown is high, rounded and open, giving little shade. The bark is the most ready means of identification.

The Bark is all colours, grey, green, bluish and brown in patches, the large scales falling entire like those of the plane tree and leaving deeply indented scars which gradually change colour. The slash is a rich yellow with a thin green bark edging.

The Wood is very dark brown. In transverse section the rings are clearly marked dark lines of varied width, rarely circular. The pores are small, in chains and festoons connected by soft tissue which shows as flecks in the dark hard tissue. The rays are fine and almost straight, invisible to the unaided eye and visible as short light bands in radial section, in which section the grain is banded and straight. In tangential section the grain is mottled, almost figured. The wood is hard, not easy to saw or plane but capable of being worked up to a rather oily finish which takes a polish. It has a not unpleasant smell. In seasoning it has been noticed that numerous radial and concentric cracks occurred but that these closed up and became quite invisible even under a lens. The weight is 50 lbs. a cubic foot.

The Leaves are pinnate, 9 inches long with an average of 11 alternate leaflets which are oval, slightly cleft, 2 inches long and 1 inch broad, increasing in size upwards, with shiny surface and short stout stalks.

The Flowers are in small racemes amongst the leaves at the twig ends. They are inconspicuous and appear from May to July. Each has 4 green sepals, 5 greenish-white petals, 10 stamens in the keel, not all fertile and a prominent, flattened pistil. They are ½ inch long.

The Fruits are pods from 2-4 inches long and an inch broad, brown with a paler edge, shiny, with veined surface and slightly embossed seeds. The 1-3 seeds are round, flat, light brown and have a white hilum, and are ⅜ inch long. The pods are persistent on the tree till the following rains.

Uses.—The wood is used for axe and hoe handles, its excessive hardness giving it the name of "kariye gatari" or "break the axe." Concoctions of the bark and roots are used medicinally.

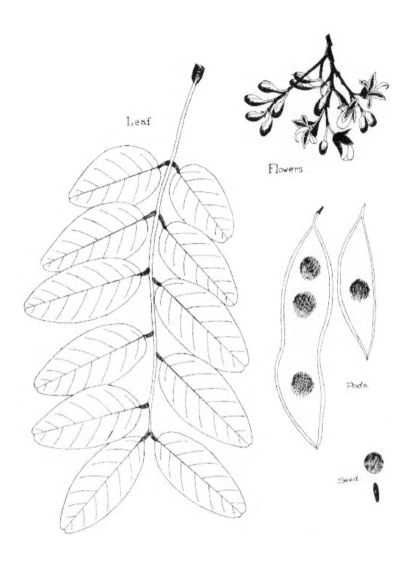

Leaf

Flowers

Pods.

Seed.

AFZELIA AFRICANA Smith.—*Kawo*. LEGUMINOSAE.

A species which is more locally than generally common and in some forests found very evenly distributed throughout large areas. Large timber specimens, as found in the south, are rarely met with, as the form of the typical northern specimens is one having a short, thick bole with a girth of 6-10 feet, large spreading limbs forming a wide, rather flat-topped crown with dense foliage and good shade. Forty feet is an average height for such full-grown trees. It is comparatively rare in the extreme north, but especially common in parts of the central and southern provinces of the north. At a distance it can be distinguished by the brilliant green of its foliage, which, though apparently dense, is actually very superficial and can be seen through in somewhat the same manner as can *Chlorophora excelsa*. It attains its full size in "kurmis" or on the slopes of well-watered hills.

The Bark is grey and flakes off in large, uneven-edged scales which leave distinctive, light patches. The slash is pale red and of a crumbling composition.

The Wood, when freshly cut is a light brown with distinct orange tint, which always remains, even when the timber has darkened to the deep red brown after seasoning. The sapwood is white or yellowish, generally the latter, the orange of the pores enhancing this. In transverse section the rings are indistinctly marked, but the soft tissue, in unusual quantities, is clearly seen, as are the pores, which are large, rather far apart, some single, or in small nests, or in lines. The rays, not visible to the naked eye, are continuous and regularly spaced, readily visible in radial section. The grain is inclined to pick up in bands, it being by no means an easy wood to work with, but the resulting finish is well worth the trouble required. It is hard, durable and strong, and weighs from 55-60 lbs. a cubic foot.

The Leaves are pinnate, about a foot long with 6-8 pairs of opposite, shiny, dark-green leaflets, 3½ inches long and 2½ inches broad. They are a most brilliant green when new.

The Flowers are in stiff, flat panicles and appear in March. Each flower has four dark-green cupped sepals, one long petal, cream-coloured with red lines, seven fertile stamens, two infertile stamens and a long, dark-green style.

The Fruits, which ripen about December or January, are large, hard, black pods, 5-6 inches long, 2½ inches broad and an inch thick, with a point at the tip. They split in half to disclose about eight large, black beans, set in a red aril in large, white cells. The pods are numerous and conspicuous on the trees.

Uses.—Planks and mortars are made locally from the wood. Furniture in small quantities has been made, and except for the weight of the wood, it is admirably suitable, in colour and texture for this purpose. The seeds are sold as a charm (fasa daga) and used in the game of "dara." The idea is held, in some parts of the country, that if the seeds be picked from the pod while it is still on the tree and made into "tuwo" the eating of them will make one safe from an attempted blow with a stick, which will remain poised, or break with the blow.

The leaves are eaten as fodder by cattle.

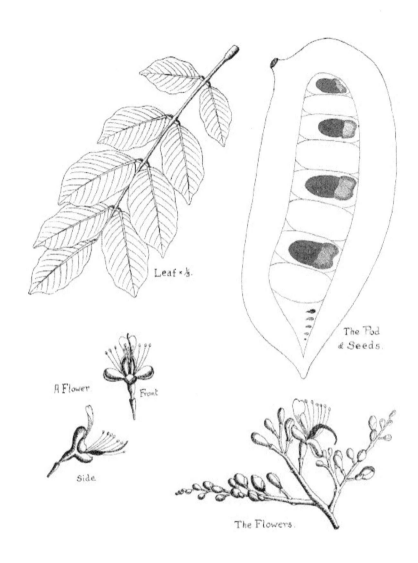

Leaf × ⅓.

The Pod & Seeds.

A Flower

Front

Side.

The Flowers.

ALBIZZIA BROWNEI Oliv.—*Madobiar rafi.* LEGUMINOSAE.

This is a large species of the tree savannah, extending generally by way of the streams and "kurmis" up to above 11° N. On suitable soils it can be found in large groups forming a high forest with the crowns meeting over 60 feet overhead. The bole length is considerable, often 30 feet and the stem is slender and straight. Enormous isolated trees inhabit the more sheltered "kurmis" and reach a height of 80 feet with a girth of 12 feet. Such specimens have small root flanges and the crown is low and of great size, not affording very much shade owing to the spreading limbs. The large bipinnate leaves of distinctive form and the big, flat pods are readily recognised.

The Bark is greyish-brown, fairly smooth, with square or rectangular close-fitting scales, larger and rougher in old trees, especially at the base. The slash is light orange in colour.

The Wood is a light red, with streaks and areas of darker and lighter red and white. In transverse section the rings are well marked and wide apart, the pores are large, mostly twins or small groups and imperfect festoons, the soft tissue not linking them up in long lengths. The rays are fine, wavy and irregular in their spacing, closing up and separating, showing as red bands in radial and fine stippling in tangential section. The wood is soft, easy to saw and plane, the finish being smooth, with a sheen and able to take a polish. The weight is 42 lbs. a cubic foot.

The Leaves are 12-18 inches long, with an average of three opposite pairs of pinnae 6-7 inches long bearing four, five and six pairs of leaflets on the lower, middle and upper pinnae respectively. The leaflets are roughly rectangular in shape due to the curving forward of the mid-rib and the blade of the leaflet being wider on the inside at the base and on the outside at the middle. The leaflets increase in size from the lowest pair to the topmost pair, varying from 1-3 inches long. They are practically the same dark, rich green above and beneath, being reddish when young. The mid-rib and nerves are prominent on both surfaces.

The Flowers appear in March and April just as the new leaves are maturing. They are of both sexes on the same tree and the same shoot, borne in small heads at the end of numerous long stalks in the leaf axils, the whole shoot being some 12 inches long. The female is solitary in the middle of the head in a circle of male flowers which open slightly later. The female is white and has a corolla of 5-6 lobes in a tubular 5-lobed calyx, and a large number of radiating styles, white with black stigmas, like a Sea Anemone in form. The male is much smaller, with a minute calyx, small tubular corolla of five points, and a long slender column of bright red stamens with black anthers radiating in a tuft at the top of the column.

The Fruits are large, flat, light mahogany brown pods, up to 8 inches long and 1½ inches broad, shining and veined, and embossed at the 9-12 seeds which are ⅜ inch long, flat, oval, sharp-edged and dark brown, with a prominent "horse-shoe" mark. The pods fall and split and the seeds remain attached alternately to the two halves till blown or washed out.

The Leaf
5 Pinnae cut off

Seed Pod
× ½

Seed.

Flowering Shoot
Leaves cut off.

ALBIZZIA CHEVALIERI Harms.—*Katsari*. LEGUMINOSAE.

A small tree up to 30 feet in height with a girth of 2-3 feet, common throughout the northern provinces. Owing to the similarity of its pods to those of *Acacia macrostachya*, it may be confused with this species in the dry season when the tree is leafless. The larger leaves, absence of thorns, corky bark and balls of flowers distinguish it at once on examination. It has an erect stem, often three or four stems from a low level, which repeatedly fork and form an open spreading crown, some of the branches extending widely and inclined to droop with the weight of the large leaves. The twigs are softly hairy.

The Bark is pale grey with a light brown tinge due to the patches left by the falling scales. It is thickly covered with very soft, corky scales, rectangular in shape and of greater length than breadth. The horizontal cuts are very straight and seem to have been done with a sharp knife. The fallen scales show a reddish colour beneath, which rapidly turns light brown and finally grey. The surface of the bark scales is very smooth and silvery.

The Wood is medium weight, rather light yellow, with a marked, long, straight grain, alternate yellow and white. The pores are very long, the rings very close. Planes and saws well, with a sheen in places.

The Leaves are about 9 inches long, bipinnate with some 10-15 pairs of pinnae, each having from 10-20 leaflets. The leaflets are broad, sharply pointed and uneven in shape, the inner portion being much narrower than the outer, whilst the inner edge is almost straight and the outer curved to the tip; they overlap and are inclined towards the tip of the pinna, whose main rib has a small extension between the top-most pair. The leaves are slightly sensitive and close at night and very shortly after they are gathered. The main rib of the leaf has purple blotches on its upper surface and is covered with very short hairs. The leaf is smooth with a waxy surface and much darker above than below.

The Flowers, which appear in April are in spherical heads of about thirty together, borne on a 2½ inch stalk. Each flower has a small 5-lobed reddish calyx, 5 small reddish petals and some 18 stamens which are an inch long, very slender, white, gradually becoming green towards the anthers. There is a 1½ inch long pistil, also becoming green towards the tip. The flowers are sweet scented, blooming at night.

The Fruits are pods from 3-6 inches long and ¾ inch wide, and vary in size according to number of seeds and in colour from a very light brown with a pinkish tinge to an uniform light red-brown colour. They are flat and the seeds embossed, the edges of the pod being straight and not indented between the seeds. The whole surface is covered with very short hairs and is velvety to the touch. The pods open on the tree or on the ground, splitting up to the stalk. The seeds are small, round, flat, and green-brown in colour, and are attached to each half of the pod alternately by a long stalk, part of which is straight and part tightly curved into an S next the seed. The pods are ripe in November and remain, a light crop, till April. There is a fine cellular veining on the surface.

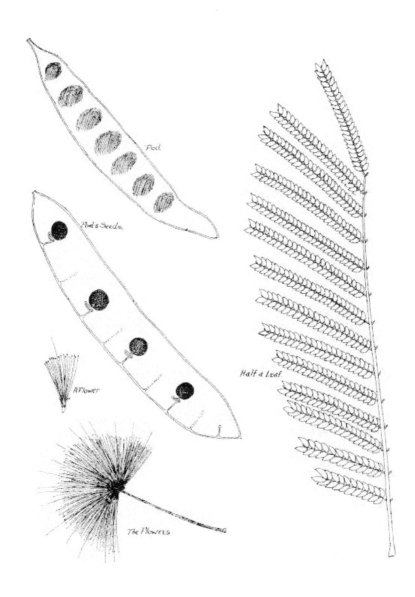

AMBLYGONOCARPUS SCHWEINFURTHII Harms.—*Tsege, Kiriya ta mata, Kolo*. LEGUMINOSAE.

This strikingly handsome tree, with a strong resemblance to the Mountain Ash, is found on poor soils as far as $12\frac{1}{2}°$ N., but in the belts where the soil is better than that of the surrounding country. Elsewhere further south it occurs throughout the higher savannah forests, often near streams, where it will grow 60 feet high, but with a small girth in proportion, not exceeding 5 feet. It averages 30 feet high, with a long clean bole and high, wide, flat crown of great beauty from the filtering of light through its bipinnate leaves whose small round leaflets are set apart, the leaves lying in the horizontal plane. A 30-feet bole on the larger trees is not uncommon. It gives little shade. It seems to be very local and where found, occurs over a small area of country in fair numbers.

The Bark is grey or brown, rough, with uneven sized scales, rounded or polygonal. These leave red scars on falling.

The Wood is a rich red-brown, more red than brown. The sapwood is grey. In transverse section the rings show as darker bands of varied width and the colour is darker in this section. The pores are numerous, in festoons of various lengths, or single, and a few nests, the whole densely and evenly distributed and the soft tissue plainly visible to the unaided eye as waves and specks in the dark hard tissue. The rays are straight, unevenly spaced, some being only the width of a ray apart, others far apart, some visible to the naked eye. The vertical section shows bands of colour and long dark pores. The wood is very hard, difficult to saw and picks up in bands under the plane, but the finished surface is smooth and will take a high polish. The weight is 60 lbs. a cubic foot.

The Leaves are bipinnate, 10 inches long, with 2-5 pairs of opposite or sub-opposite pinnae bearing 8-16 alternate leaflets, broad oval, flat tipped, pale blue-green, on light brown stalks, each ¾ inch long and ½ inch broad. They spring from the ends of erect twigs and lie in the horizontal plane. The leaflets are set apart and do not touch or overlap.

The Flowers, which appear in March, are in spikes, 2-3 inches long, crowded with white, scented flowers. Each flower has five small white petals with acute tips, ten white stamens with yellow anthers and a white pistil. They are not very conspicuous as they are partly hidden amongst the foliage.

The Fruits are large four-angled pods, the suture angle rounded. They are 4 inches long and 1½ inches broad, brown, pendulous on a 3 inch long stalk. They contain 6-10 seeds, roughly four-angled, pointed at one end. These lie across the length of the pod and ripen about December. They are ¾ inch long.

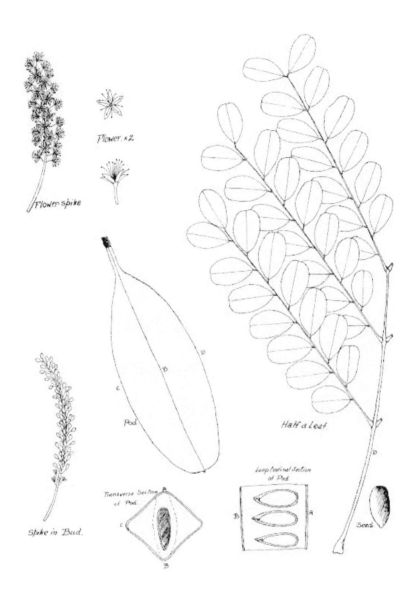

Flower x 2.

Flower-spike

Flower-spike

Pod.

Half a Leaf.

Spike in Bud.

Transverse Section of Pod.

Longitudinal Section of Pod.

Seed.

ANDIRA INERMIS H. B. K.—*Gwaska, Madobia.* LEGUMINOSAE.

A large tree occurring near streams in Bauchi, Sokoto and Zaria provinces. It is also found in the forests where the conditions are moist enough, and is not at all uncommon amongst the rocks on the Bauchi plateau. It rarely has any length of bole, the crown almost reaching the ground and the stem being bent or leaning. The crown is dense and the branches droop at the end. A height of 30-40 feet with a girth of 4-5 feet is usual. The foliage much resembles that of *Khaya senegalensis*, from a distance, but has a brilliant sheen.

The Bark is brown or grey-brown, roughish, with scales an inch or two in diameter fitting close together.

The Leaves are pinnate, 10-18 inches long with 5-7 pairs of opposite or nearly opposite leaflets and a terminal leaf. The smallest at the base are 1½ inches long and ¾ inch wide; the largest in the middle are 3½ inches long and 1½ inches wide. The edges are parallel and there is a sharp taper to a cleft tip. The basal lobes are uneven and the leaflets vary much in shape. The terminal leaflet is broader in proportion to its length than the lateral leaflets and the margins are not parallel. There are small stipules at the base of each stalk. The surface on both sides is shining and the new leaves are a most brilliant green. The mid-rib is below the upper surface but raised beneath.

The Flowers which bloom from March to May are in loose panicles up to 2 feet in length and in such numbers as to show up from a considerable distance. The blossom is very ornamental. The flower is papilionaceous with a large pink standard and white keel. A species of small black ant swarms over the flowers at times.

The Fruits resemble large walnuts and hang in bunches on long stalks like mangoes. They are ovoid, 2-3 inches long and 2 inches across, with a thick, fibrous, green-brown case containing two white kernels one above the other. They ripen about September. There is the line of a suture down one side, and the surface is uneven.

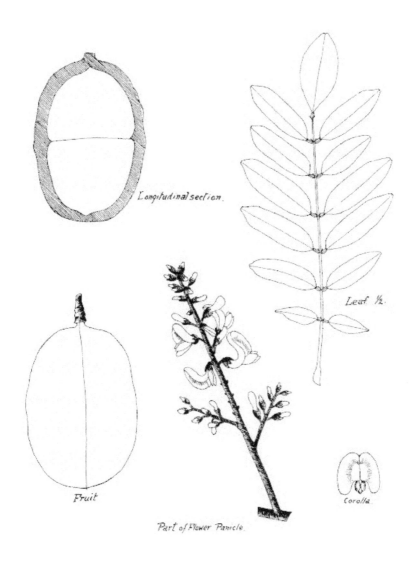

Longitudinal section.

Leaf ½.

Fruit

Corolla

Part of Flower Panicle.

ANOGEISSUS LEIOCARPUS Guill. & Perr.—*Marike.*
COMBRETACEAE.

This is perhaps the most evenly and widely distributed of all trees, extending from the southern rain belt to the extreme north of Nigeria, and over the border into French territory north of Sokoto, where some of the largest examples of the northern provinces are to be found. A height of 70 feet with girths of 6-8 feet is common. As regards the situations most favourable to this species, level country with deep soil and average rainfall is preferred, and considerable areas of such country are found occupied almost entirely by it, both as full grown trees and saplings in dense thickets. These latter are especially valuable for the production of forked poles (gofa) for building purposes, and are transported great distances owing to their great durability and comparative resistance to White Ants. The species is readily distinguished at a distance by its feathery, birch-like foliage and drooping branches, which, at first acutely ascending, bend over and droop in graceful curves. A light-demander, its branches are extremely persistent on the stem and only where it is very densely grown, preferably with an admixture of shade-bearing species, will clean boles be found. Examples grown in the open, or suddenly freed from the forest by the clearing for cultivation, are covered with small twigs and the foliage extends almost to ground level. Large specimens in forest have clean boles, generally forking at no great height, with wide-spreading, open crowns.

The Bark varies according to habitat. That of specimens in dense forest is fairly smooth with small, brown scales, but as generally seen in the north is a very light grey with long scales which turn up at the ends before falling, giving the tree a ragged appearance, with large, lighter patches. A dark gum exudes from the bark, very inferior as an adhesive as it is cloudy when made up into solution. The slash is pale yellow with thin dark lines.

The Wood is a dark, dull, smoky brown, sometimes almost black, with reddish streaks. The sapwood is grey or dirty white. In transverse section the rings are seen as dark bands, the pores are minute and densely distributed in small groups, chains and festoons of various sizes, with single pores between, the soft tissue sparse and hardly connecting the pores. The rays are extremely fine and close together. In vertical section the grain is very fine and the pores have glistening contents. It is a very hard wood, difficult to work with all tools, often knotty, picking up in bands, but will polish, even on the transverse section. The weight is 64 lbs. a cubic foot.

The Leaves are on long, slender, drooping twigs, and are oval, slightly darker above than below with a silky pubescence which is most marked in the young foliage. They tend to assume one plane. Length 2-3 inches, breadth 1 inch.

The Flowers are little cream-coloured, scented balls, ⅜ inch diam., from February on. Each has a 5-pointed calyx, 10 erect stamens, a short, straight pistil, round the base of which is a ring of reddish hairs. They are very inconspicuous.

The Fruits are small, rough, cone-like balls of irregularly serrated-edged seeds, packed horizontally. They are green in the seed and brown in the wing portion. They ripen about December and are about ¾ inch.

Uses.—The wood is burnt entire and the fine, white ash used for fixing dyes. The ash is also used for washing white garments. The roots are used as

"Chew-sticks" for cleansing the mouth and teeth. The saplings and branches produce forked poles (gofa) for building. The bark yields a dark, inferior gum. The seeds are a cure for worms in horses. The leaves, mixed with salt, make a yellow dye.

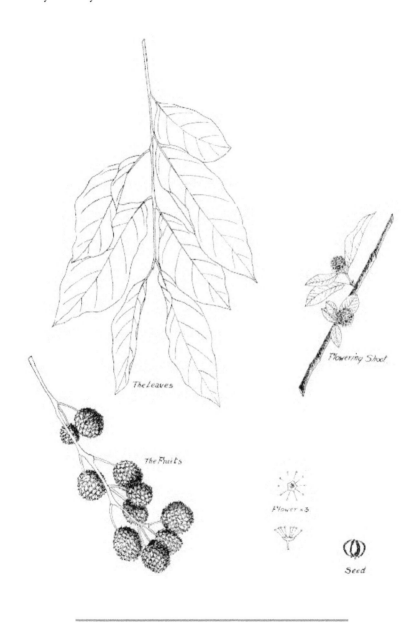

The Leaves

Flowering Shoot

The Fruits

Flower x 3

Seed

ANONA SENEGALENSIS Pers.—*Gwandar daji.* *"Custard Apple."* ANONACEAE.

A very common shrub, averaging some 10-12 feet in height, sometimes a small tree, in suitable situations exceeding 20 feet; very widely distributed and found everywhere except in the extreme north. The distinguishing features are the oval, bluish leaves, the waxy, yellow flowers and the familiar orange-coloured fruits. It has no particular form.

The Bark is normally light, silvery grey, smooth, with marked horizontal cuts round the stem, having the appearance of joints. Older or exposed stems, especially those subject to fire, become darker and roughly scaly. The slash is a dull, pale pink.

The Wood is greenish-grey. The transverse section shows indistinct rings, minute single pores scattered in between the fine waved rays which reach from the centre to the edge of the wood and show as small dark bands in radial section. In vertical section the grain shows as slight variations in colour. It is soft and easy to saw and plane. Weight 40 lbs. a cubic foot.

The Leaves are oval, some 6 inches long and 3 inches broad, bluish-green, with a short stalk, the lateral nerves inclined well forward and the small connecting veins parallel, numerous and at right angles to the mid-rib. The leaves stand erect and are apt to fold up along the mid-rib. When crushed they are fragrant.

The Flowers appear from January to April, single or in pairs, ¾ inch long, pale yellow and waxy. Each has three small green sepals and six petals in two rows, the outer three large, with broad flat edges which meet tight together in bud, the inner three smaller, thick and pale yellow, their tips meeting over a mass of stamens round the ovary and stigmas. The flowers do not open wide and last for a long time.

The Fruits are rounded, 2-3 inches long, composed of fleshy carpels each containing a seed, green at first with a resemblance to cones, ripening to a rich orange colour. They are pleasant but unsatisfactory eating owing to the large number of seeds embedded in the juicy flesh. Birds are very partial to them, hollowing them out before they ripen.

Uses.—The fruit is eaten fresh.

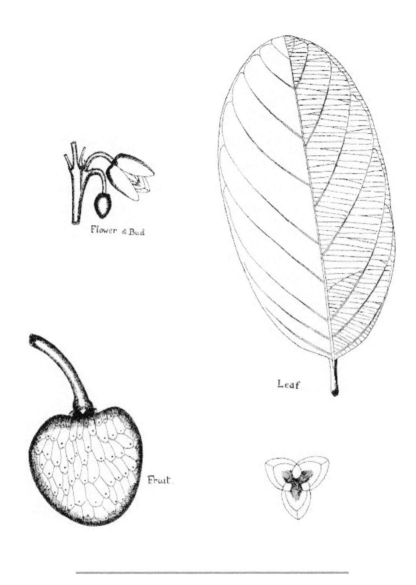

Flower & Bud

Leaf

Fruit

BALANITES AEGYPTIACA Del.—*Aduwa.* *"Desert Date."*
SIMARUBACEAE.

In the north this is one of the commonest trees on loose sand or barren, stony wastes, where it is often the only species to survive the extreme conditions. It averages 15-20 feet high, but 30 feet is not uncommon, with a girth of 4-5 feet. The long green thorns and small dark leaves distinguish it. The form is roughly spherical, with a tangled mass of long thorny twigs, whose ends droop or protrude here and there some feet from the main thicket. It extends northwards to the limits of tree growth, affording fodder for camels and goats.

The Bark is grey, with long, wide, deep, vertical fissures in which the yellow of the new bark can be seen. The scales are long, thick, prominent and ragged. The branch bark is distinctive. Dark green and smooth, the cream-coloured lenticels of various lengths cover it ever more thickly from the tip downwards till the green colour is completely obscured and the grey of the branch bark is reached. The slash is pale yellow.

The Thorns are modified shoots borne spirally round the long slender twigs in the leaf axils, 2-3 inches long, tapering evenly to a strong, sharp point. They incline slightly forward and barely curve. They are dark green and persist on the branches.

The Wood is light yellow with sometimes greyish discolorations. In transverse section the rings are distinct in wavy light lines; the pores are small, open and in little groups in concentric rings and the rays are clearly visible as light lines of various overlapping lengths. In radial section the rays add a figure to the wood in lines up to an inch long. In vertical section the wood is of even colour and texture, close grained and sound. It is easy to carpenter and planes smoothly without picking up. Weight 48-50 lbs. a cubic foot.

The Leaves, with the thorns in their axils, are really paired leaflets on a short, common stalk; 1½-2 inches long, over 1 inch wide, a very dull, dark green, they are unequal lobed, as illustrated, with a tendency to close up along the mid-rib. They assume a vertical position, even when the twig droops, the stalks bending to adjust themselves to this position.

The Flowers, in the leaf axils, are found from November to March, in spikes up to 3 inches long, often shortened to resemble round clusters. Each flower is ½ inch across, with five small green sepals, five longer and darker green petals, ten short yellow stamens and a shining dark green ovary with short, blunt pistil. The tree sometimes flowers out of season.

The Fruits, which ripen from February onwards, are 1¼ inches long and about an inch wide on a short stalk. At first green with a wrinkled, nipple-like tip, they turn yellow on ripening, when they have a thin hard skin, a light brown, sticky, edible flesh and a large, hard, pointed stone. There is a space between the flesh and the skin.

Uses.—The fruits are edible, particularly appreciated by beasts. The wood makes excellent axe and hoe handles. Planes have been made of it, it being similar to beech in quality. The branches are used for hedging farms.

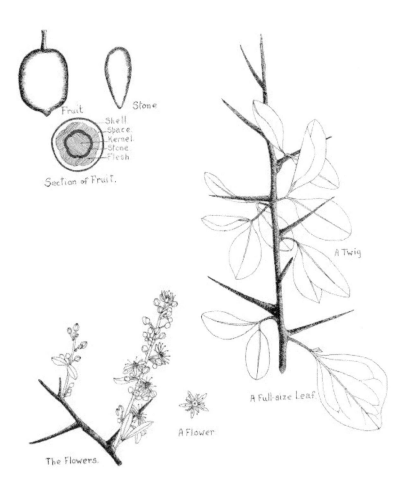

Fruit

Stone

Shell
Space
Kernel
Stone
Flesh

Section of Fruit.

A Twig

A Full-size Leaf.

A Flower

The Flowers.

BALSAMODENDRON AFRICANUM A. Rich.—*Dashi.*
BURSERACEAE.

This shrub or small tree is locally plentiful and is found in all situations, in the driest sandy soils, on the tops of rocks where it grows stout and stunted in the crevices, and in loamy soils. It is frequently seen planted as a live hedge from cuttings and forms an impenetrable barrier after the habit of the Blackthorn. The slender but rigid twigs shoot out in all directions and with its trifoliate leaves on the purple twigs, the plant is readily identified. As a small tree, a form met with on hills or in better soils in the forest, the stem is short and stout, dividing low down and quickly branching into a light rounded crown. The height is rarely over 15-20 feet.

The Bark of old examples is green and shining, covered with little papery scales which flake off. That of the branches and twigs is red or purple. A resinous gum with a pleasant scent exudes from the slash.

The Thorns are branches whose tips are modified in the form of spines, which will bear leaves, and in the dormant season are dotted with buds.

The Leaves are trifoliate, often very small, but when full-grown some 3-3½ inches long, of which one inch is stalk, red above, green beneath. The middle leaflet is broadest slightly over half-way up, tapering gradually to the base and more suddenly to the tip, with irregularly serrated edges. The lateral leaflets are unequal in size, the right hand one usually considerably larger than the left, both rounded and serrated in the same manner as the middle leaflet. The surface is shiny waxy green, very bright when young. The venation is prominent beneath, the lateral nerves much branched and zig-zag. The leaf is scented.

The Flowers appear in October on the leafless tree and are in small clusters all along the twigs, up to about 10 in a cluster, each on a minute stalk. The flower is ¼ inch long with a 4-lobed tubular corolla deep red with green lobes, the petals separate but overlapping and close together in the tubular portion which is held in the 4-lobed, cup-shaped calyx. There are 8 stamens, 4 just appearing in the corolla mouth, 4 shorter, and a pistil wholly below the corolla mouth.

The Fruits are small irregularly pear-shaped drupes a little over ¼ inch long, greyish with a purple bloom which is easily rubbed off. They have a highly resinous flesh and white kernel and ripen in the early spring. They grow in clusters along the twigs and the crop is often a heavy one.

Uses.—The resin is used as a scent on garments and medicinally, taken internally.

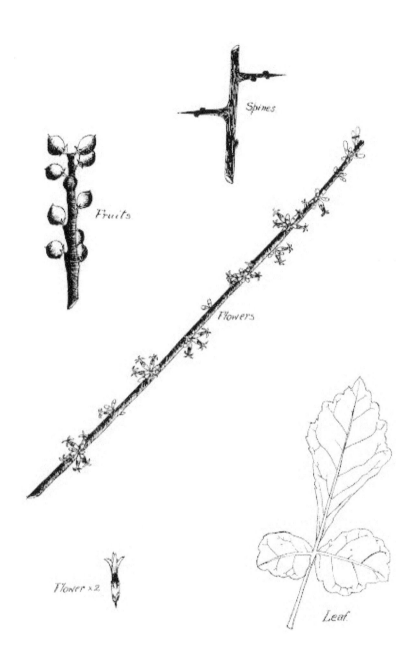

Spines

Fruits

Flowers

Flower x 2

Leaf

BAUHINIA RETICULATA DC.—*Kalgo, Kargo*. LEGUMINOSAE.

A medium-sized tree or shrub, in which latter form it occurs over large areas of country as pure growth, particularly as secondary growth in farmed land, where it is a great pest to the farmer who desires to continue sowings. It grows prolifically from stumps which the farmer does not trouble to uproot. It is very fond of small, shallow depressions, where it will grow to the exclusion of all else. In its proper habitat it will grow over 30 feet high with a girth of 8-10 feet, with short bole and an enormous rounded crown of dense foliage affording good shade. It is one of the commonest species in the north.

The Bark is a dull, dark grey, sometimes with a rust-red tinge, deeply fissured and ridged with hard, brittle bark of some thickness, which falls in large, ragged sections. That of young trees can be ripped off readily after being cut. The slash is bright crimson, turning brown on exposure, and shows the fibrous nature of the bark.

The Wood is oak-brown, rather dirty looking, with light patches and dark discolorations, especially round flaws. In transverse section the rings are indistinct but the hard and soft tissue is very well marked, mostly in concentric lines. The pores are small, evenly distributed, mostly in festoons joined by the soft tissue. The rays are extremely fine and closely spaced, invisible to the unaided eye and very faintly seen in radial section as small bands. The wood is easily sawn but picks up badly under the plane. It is very strong and tough and weighs 50 lbs. a cubic foot.

The Leaves are bifoliate, the depth to which the leaf is divided varying a great deal and the angle being sharp or rounded. They average 4 inches across but those from stool shoots especially may be very much larger and generally the larger the tree the smaller the leaves. There are four main nerves on each leaflet and the mid-rib projects slightly between them. The leaf stalk is an inch long, with large base. Like all species of this genus the leaf tends to fold up along the mid-rib. The colour is lighter beneath and the venation a very intricate network and prominent. The texture is tough.

The Flowers are in spikes, 2-6 inches long and appear in February or March or even as late as June. Each has a 5-lobed calyx, green-brown in colour, 5 white petals, wrinkled and overlapping at the edges, 10 stamens of varied length with brown anthers and a short, blunt pistil. The flowers drop off readily when handled and are about an inch long. They do not open very wide.

The Fruits are pods 6-9 inches long, hard, flat, dark-brown, straight or contorted into strange shapes, 2 inches broad and ¼-½ inch thick. They are persistent on the tree for many months, a most disfiguring feature, and drop entire, rotting on the ground. They are very liable to attacks from a grub which destroys the seeds. These are small, brown, oval and scattered about in the mealy endocarp which has an objectionable smell.

Uses.—The wood is used for axe and hoe handles of all sorts. The bark is used for binding and tying but is not plaited or twisted into ropes. Cattle eat the pods.

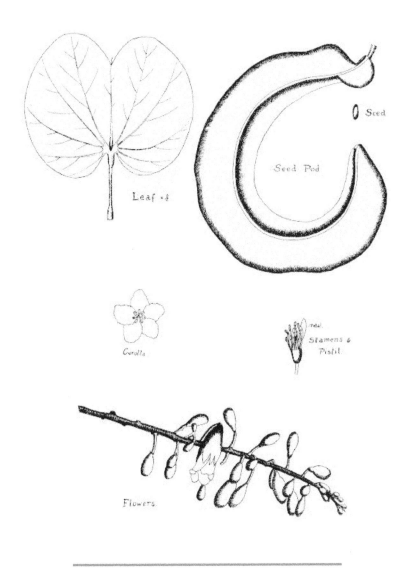

Leaf ×¼

Seed

Seed Pod

Corolla

Petal.
Stamens &
Pistil.

Flowers.

BAUHINIA RUFESCENS Lam.—*Jirga, Tsatsagi, Matsagi.*
LEGUMINOSAE.

A small tree, typically inhabiting annually inundated areas but also flourishing in sandy soils of poor quality. It forms, with its numerous much branched stems, a small thicket of one or more trees; it is difficult to say how many under the circumstances, as one root-stock will produce a number at ground level, forming an impenetrable mass. It can be distinguished readily by the typical leaves of the genus, and the small size of these compared with other species. The fact that it is always in flower and fruit will identify it also.

The Bark is a light ash-coloured grey, smooth and covered with small, brown horizontal lenticels. Old stems bear at the base small dark scales and the bark is fissured giving it a speckled appearance. The slash is pink and reveals the very fibrous nature of the bark.

The Thorns are modified twigs whose tips are strongly pointed. They may support leafy shoots or bear leaves themselves and are seen best on the long, slender shoots and drooping twigs. They extend their growth indefinitely, but on the older wood may be seen as bare, woody spines 3 or 4 inches long, curved out and down.

The Wood is a dull, smoke-brown colour. In transverse section the rings are indistinct dark lines; the pores are numerous and unevenly distributed, single in small groups or in short chains, the rays invisible to the unaided eye but showing as small bands in radial section, which reflect the light. In vertical section the grain is close and there are darker bands of colour. The wood is not hard, is easily sawn and planes to a nice, soft, smooth finish though it picks up in places with the long soft fibres. The weight is 50 lbs. a cubic foot.

The Leaves are borne spirally round the branches, often in the thorn axils or on the thorns themselves, or on twigs bare of thorns. They are bifoliate, typical of the genus, rarely over an inch long and mostly a ½ inch, the terminal leaf of a new shoot often being much larger. The mid-rib is extended between the leaflets which are divided to the base and there are three main nerves to each leaflet. They are a light, blue-green or grey-green, with a bloom and are partly sensitive, closing up rapidly after being plucked.

The Flowers are white and can be found all the year round, though the proper flowering season is in the rains or towards the end of them. They are in small racemes at the twig ends, pendulous amongst the leaves, ¾ inch long, with a calyx which splits partially into five parts, remaining attached at the tips and opening beneath, 5 white petals, spoon-shaped with narrow base and broad, pointed tip, 10 white stamens with light brown anthers and a tuft of hairs at the base of each, and a clubbed pistil.

The Fruits are pods about 3 inches long, constricted between the seeds, dull black, slightly curved, containing up to eight shiny, rich red-brown seeds, roughly rectangular with one rounded corner. The seeds rattle loose in the pod which falls entire and rots on the ground. The pods hang in conspicuous clusters and may be seen all the year round.

Uses.—The bark is stripped for binding but is not plaited into ropes.

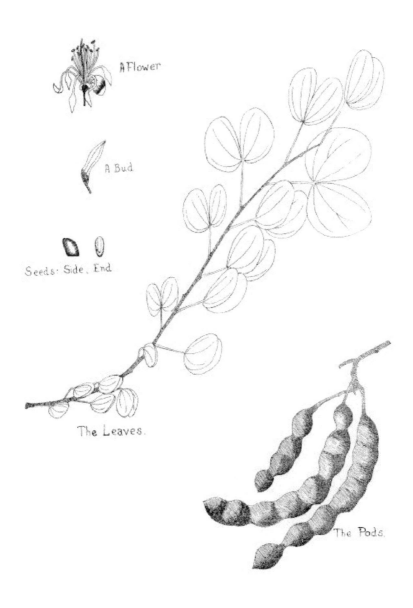

A Flower

A Bud

Seeds: Side, End

The Leaves.

The Pods.

BERLINIA AURICULATA Benth.—*Doka rafi.* LEGUMINOSAE.

This is a medium-sized tree of the best savannah types, commonly found in "kurmis" and extending as far north as 11°. It occurs plentifully in Sokoto and Kontagora along the small streams in the south. Averaging 40 feet in height it will reach 60 feet with girths of 8-10 feet. The crown is high, rounded and very dense in the open; flatter and more superficial in heavy forest, topping a 30-feet bole, only 10 feet long in open situations. The flowers and pods are both conspicuous.

The Bark is dark-grey or brown, the scales leaving large, concave scars. The slash is pale brown.

The Wood.—The heartwood is red-brown, the sapwood light with a pink tint. In transverse section the rings are well but unevenly marked red lines. The pores are large and solitary, for the most part in oblique rows, and the rays are extremely fine and invisible to the unaided eye. In vertical section the pores are long and straight and the rings show as dark bands. It is rather a coarse, stringy wood, sawing roughly and picking up in long fibres under the plane. Where the grain suits the plane the finish has a bronze sheen. It is a fairly hard wood and weighs 50 lbs. a cubic foot.

The Leaves are pinnate, 9-12 inches long with an average of four pairs of pinnae increasing in size from the lower pair upwards, the lower 2-4 inches long and 1½—2 inches broad, the upper pair 5-7 inches long and 2-3 inches broad. The mid-rib curves forward, rounding the outer edge of the leaf. The nerves are alternate long and short. The similarity of the leaf and fruit to those of *Isoberlinia doka* gives it its native name.

The Flowers are in close panicled racemes at the branch tips from March to June. Each flower is 4 inches long over all, being enclosed at first in a pair of long, pale green, velvety bracteoles which separate and fall back to release the flower parts. The calyx is a slender tube divided into five long, narrow pointed, recurved sepals. The corolla has four small, linear petals and one erect white, wrinkled, cleft petal 2 inches long and broad, with a green centre splash. There are ten long, erect, hairy stamens with brown-green anthers, and a long pistil. The flowers are in such masses as to be conspicuous from a great distance.

The Fruit is a broad, flat, dark brown, velvety pod, 6-10 inches long and 2-2½ inches broad. It explodes when ripe to release some 3-5 round, flattened seeds.

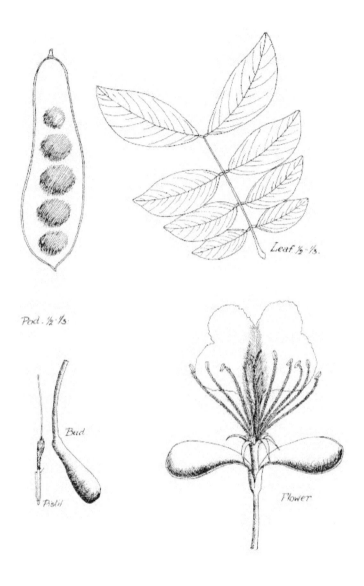

Pod. ½-⅓.

Leaf ½-⅓.

Bud

Pistil

Flower

BOMBAX BUONOPOZENSE Beauv.—*Kuriya, Gurjiya.* MALVACEAE.

This tree is known as the Red-Flowered Silk Cotton Tree and grows, on an average, to a height of 40 feet with a girth of about 5 feet. Very much larger specimens are, however, to be seen, up to 70 feet high with girths of 15 feet and more. Rounded root-flanges are prominent on the large trees. The crown is regular and umbrella-shaped, rather flat-topped, wide-spreading and open with superficial foliage giving little shade. In old specimens the bole is gnarled and the limbs much bent, the shape of the crown being often retained by the growth of the smaller branches. The species is not at all exacting as regards soil and is, in fact, commonly found on hillsides and amongst rocks where its roots penetrate the little but good soil. It is more local than general in its distribution.

The Bark of the younger trees is very corky, longitudinally fissured and horizontally cut into prominently spiked scales. The degree of roughness is very variable, being generally far more marked, relatively on younger trees, and, in the older ones, the thorns being confined to the branches, the bole having prominent layered corky scales with soft ends. Some quite small trees bear no thorns at all, but this is rare. The spines are conical, with broad corky bases and sharp black points. If a piece of the bark be snapped off it shows a light red colour. The slash is crimson.

The Wood is a dirty white colour. In transverse section the pores are large, widely scattered, single, twin or nests. The rays vary in width and in spacing very considerably, and are straight, showing as long, light-reflecting bands in radial, and as brown flecks in tangential section. In the latter section the pores are brown, open and wavy. It is a very soft and light wood, easily worked, not strong, its durability largely dependant on the manner in which it has been seasoned, with exclusion of damp and consequent mould. It is very subject to small borer beetles. The weight is only 20 lbs. a cubic foot.

The Leaves are truly digitate with generally six lobes, broad at the tip, narrow at the base, with a prominent point. The entire leaf is 6-7 inches across, with a 5 inch stalk. The venation is prominent on both surfaces and the colour is pale, the surface smooth.

The Flowers are the most conspicuous feature of all and may be found from November to February. They are red, tulip-like blooms, 3 inches in diameter, with a dark red 5-lobed calyx, 5 red petals and a mass of black-anthered stamens filling the corolla and surrounding a 5-part pistil. They fall in numbers, entire, and are devoured by antelopes.

The Fruits are large, pendulous capsules some 4 inches long and 2 inches wide, black or deep brown when ripe, splitting into five sections to release small black seeds embedded in a mass of silk-cotton. The pod shrinks in the ripening and the silk cotton is packed tight in it and expands in bulk enormously by hygroscopic action, carrying the seeds a great distance on the wind.

Uses.—The wood is used for making large and small drums, native stools used by women, basins, shoes and saddles and cattle troughs. The bark is used by women to impart a red colour to the teeth. Certain pagans make a sauce from the flowers, locally called "Kwungi."

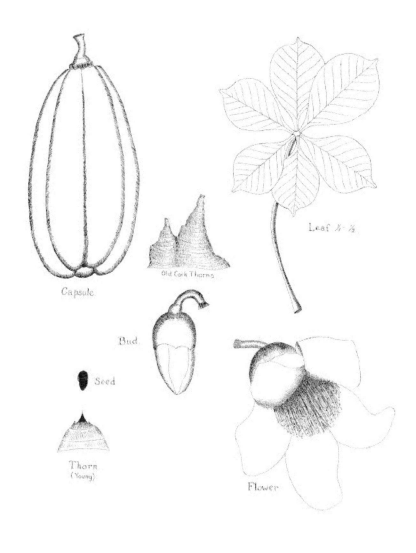

Capsule.

Old Cork Thorns

Leaf ⅓-½

Bud.

Seed

Thorn
(Young)

Flower

BORASSUS FLABELLIFER Linn. *var.* **AETHIOPUM** Warb.—*Giginya.*
PALMACEAE.

This very well-known Palm inhabits marshes, the banks of streams and any well-watered hollow, and will grow, if introduced, on dry sandy soils. With the exception of the Dum Palm, *Hyphaene Thebaica*, it is the most northerly of the palms in Nigeria and is found in vast quantities in "fadamma." It does not actually grow in the water, though it will survive an occasional immersion when grown, but occupies islands, banks and edges of marshy land as well as the banks of streams, lakes, etc. It reaches a height of over 80 feet, with a girth at the base of 6 feet, at breast height 4 feet 6 inches, and at the narrowest point 3 feet. The stem swells at about 30 feet and after the palm is about 50 years old narrows again, repeating the swelling process again and even a third time in very old palms. The rate of growth, except when cultivated, in which case it is quicker, is very slow. The seed germinates in a month and sends down a shoot from 2-4 feet deep into the ground. This shoot swells at the base and loses connection with the empty seed, and sends up a green shoot from the base, which forms the first leaf. In the root-bud stage it is eaten as a vegetable, called "muruchi" and thousands of seeds are planted for this purpose alone. The first leaf is a narrow blade, as are subsequent leaves, gradually broadening till the first frond appears about three years later, the stem appearing at ground level after some six or seven years. From then onward the growth is at the rate of from 12-18 inches a year, 50 year old palms being about 50-60 feet in height.

The Bark.—The stem is a smooth one after a number of years; the leaf scars, very marked when fresh, gradually fading away.

The Wood is heavy and hard, apt to splinter into separate fibres, but very durable above ground. It weighs about 50 lbs. a cubic foot. The wood of the male is more compact throughout than that of the female whose centre is looser, the outside only being really serviceable.

The Leaves are up to 12 feet long, fan-shaped, the segments V-shaped, joined for half their length, the stalk concave above and spiny on the margins. The sheath divides at the base before falling and remains for some time clasping the stem.

The Flowers appear in April, just as the fruit has fallen, the females on one tree the males on another. The males are on branched spadices up to 6 feet long, the separate catkins a foot long and 2 inches broad. There are 3 sepals, 3 petals and 6 stamens, the flowers being green. The females are on a 6-8 feet long spadix, unbranched, with up to a dozen spikes of flowers, with 3 sepals, 3 smaller petals and the rudiments of 6 stamens.

The Fruits are the most conspicuous feature, ripening in April to a rich orange colour. They are some 6 inches long and 5 inches broad, the enlarged calyx cupping the fruit. The edible, fibrous pulp surrounds three seeds with hard fibrous coats and edible kernels.

Uses.—Locally the pulp is eaten raw or pounded with milk. The kernel is eaten young. The root-buds are roasted and are a delicacy. The leaves are used for mats, bags, baskets, fans, &c. Some tribes extract salt from them. The wood is used for canoes, rafters, poles, water-pipes, doors and guttering. Elsewhere the leaf stalks are the source of fibre, a spirit has been distilled from the flower spathes and buttons made from the seeds.

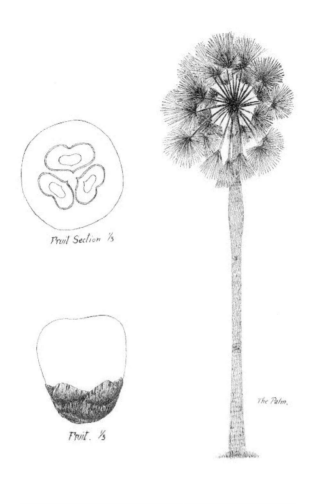

Fruit Section ⅓

Fruit. ⅓

The Palm.

BOSWELLIA DALZIELII Hutch.—*Hano, Ararabi, Basamu.*
BURSERACEAE.

A very common tree in high savannah, extending as far north as 13° where the soil is suitably rich and well watered. It will not grow as far north as this where the conditions are dry and barren, granite soils and situations on stony ground of the right kind being suitable. In sand or laterite it will not thrive, but it will flourish on bare granite where its roots can make their way into crevices. In some suitable localities it forms almost pure forest and is a very handsome species under those circumstances. The leaf, flower, fruit or bark are all distinctive. Another closely allied species, *B. odorata*, differs only in having a branched panicle in place of the bunch of racemes of *B. Dalzielii*. When young or middle-aged the stem is erect and the limbs ascend steeply forming a high crown, foliated down to a low point. Older trees exhibit a short, massive bole from which the heavy limbs spread out forming huge crowns with drooping extremities.

The Bark is most conspicuous, pale brown, with large papery pieces peeling off and at times hanging in shreds from the stem. The slash is reddish-brown and a scented gum-resin exudes, partly drying into nodules, almost white in colour, readily crumbled.

The Leaves are 12-18 inches long, pinnate, with some 7-9 pairs of long, slender, pointed, deeply-toothed leaflets. These increase in size towards the top end of the leaf, the basal pair often being very small and distinct in shape. The terminal pair are frequently partly united into one for half or more of their length. The leaflets are sessile, in colour light green and shining, with the venation raised on both surfaces.

The Flowers appear from January to April, and are in large bunches of racemes at the tips of the large, blunt twigs. The racemes are from 6-8 inches long and bear numerous white flowers on $\frac{1}{2}$-$\frac{3}{4}$ inch long stalks. Each flower is $\frac{5}{8}$ inch in diameter with 5 white, pointed petals, 10 stamens whose anthers are bent away from the filament at a side angle, and a small, blunt pistil. The flower disc is red and the flower scented.

The Fruits are 3-angled capsules, elongated pear-shaped, with prominent bulges opposite the seeds. They are $\frac{3}{4}$ inch long and the stigma remains are present. The capsule splits into three and the small seeds are released, each with a sharp spike at the top end.

Uses.—The gum-resin, being scented, is used for burning in houses, or fumigating clothes, and as an ingredient in medicine.

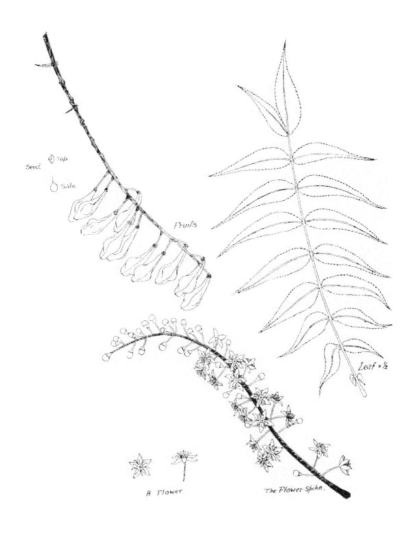

Seed
⊙ Top

◯ Side

Fruits

Leaf ×½

A Flower

The Flower Spike.

BRIDELIA FERRUGINEA Benth.—*Kisni, Kirni*. EUPHORBIACEAE.

A small tree of the Tree and Bush savannahs which does not extend very far north and will not grow away from fairly good soils. It is rather shapeless, except occasionally in closer forest, where it may reach a height of 30 feet with a girth of 3-4 feet. This species and *B. scleroneura* are similar in habit, with open, formless crowns, straggling, drooping limbs covered with numerous, slender, erect twigs which are apt to be burnt back repeatedly by fires and show the position of leaves and flowers as prominent clusters at frequent intervals. The distinguishing features are larger leaves of this species and the absence of the marked crenate edges.

The Bark is dark grey, covered with small, close-fitting, prominent scales. That of the branches is lighter and covered with rectangular scales, clearly cut in rows. That of the smaller branches is almost white with vertical, wavy ridges of cork. The slash is crimson.

The Wood is brown and the sapwood is dull white. In transverse section the rings are clear, fine light lines, the pores are very small, single, double, or in small festoons, the soft tissue not well developed. The rays are fine, close, fairly straight and evenly spaced. In vertical section the grain is coarse, and not very straight and the wood is easily sawn but hard to plane, picking up in patches. The weight is 60 lbs. a cubic foot.

The Leaves are alternate, assuming one plane. They are 4-5 inches long and 2¼-2½ inches broad, with slightly waved edges, the nerves reaching the edge. They are dark green above, olive green beneath, with the venation very prominent and covered with brownish hairs, especially on the short stalk and mid-rib.

The Flowers, male and female on the same tree are in small, compact clusters in the leaf axils from March to June. The male flowers have a calyx of 5 sepals, 5 minute petals and 5 radiating stamens, joined at the base into a short column. The female flower has the same calyx, longer, narrower petals and two styles which are forked. The male flowers have a shiny yellow disc.

The Fruits are small black ovoid drupes, becoming veined and wrinkled when dry, with a single cell containing two seeds. They are ⅜ inch long and flattened at the top, and very persistent on the twigs.

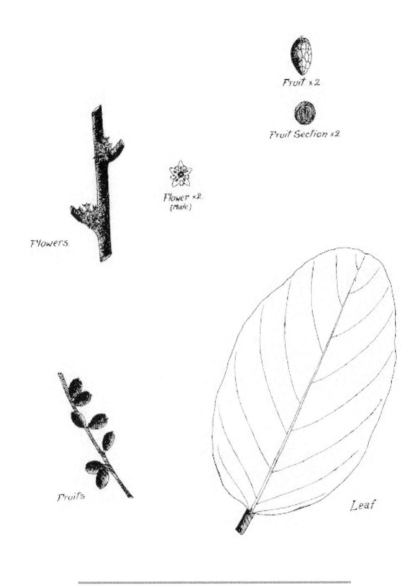

Fruit x2

Fruit Section x2

Flower x2
(Male)

Flowers

Fruits

Leaf

BRIDELIA SCLERONEURA Müll. Arg.—*Kisni*. EUPHORBIACEAE.

This is a small tree, without any definite form, which grows up to 15 or 20 feet high, with, occasionally a stem some 6 feet high and a misshapen crown. It is distinguished by the small clusters of minute flowers in the leaf axils and the crenate margined leaves which lie in one plane on long, slender, drooping twigs.

The Bark, in old trees, is quite black, with long fissures and fibrous ridges; in young trees it is white, or black and white in patches, with soft, corky scales which fall in large pieces composed of several smaller scales bound together. The slash is bright crimson.

The Leaves are borne on long, drooping, slender twigs, alternate, in one plane and gradually diminishing in size from the base upwards. Each such twig is supported by a larger leaf, 3½ inches long and 1¼ inches broad. The alternate leaves vary from 3 inches long and ¾ inches broad at the base of the twig to 2 inches long and ½ inch broad at the end of the twig. The lateral veins, as they reach the margin, are continued along it so that the edge is crenate. The tips are sharply pointed. The leaves all point well forward. The ¼ inch long stalks are covered with soft, brown hairs. The upper surface is slightly rough and darker than the under surface, on which the veins, especially the laterals and mid-rib, stand out prominently and show yellow in colour. The halves of the leaf tend to fold up along the mid-rib slightly.

The Flowers are in very small clusters in the leaf axils and appear in May and June. They are bright red in bud, the sepals being tinged. Each male flower is ⅛ inch in diameter with a 5 sepal calyx, 5 minute petals and a column of 5 radiating stamens. The female flower has 5 sepals, 5 petals, a disk round the ovary and 2 bifid styles.

The Fruits are in clusters, often a dense crop, at each node. They are the size of peas, green at first, ripening to black with a bloom. A thin skin surrounds a narrow, juicy flesh round a hard, 2-celled stone. The flesh dries and the skin wrinkles while the fruit is on the tree.

Uses.—The roots are sometimes used medicinally.

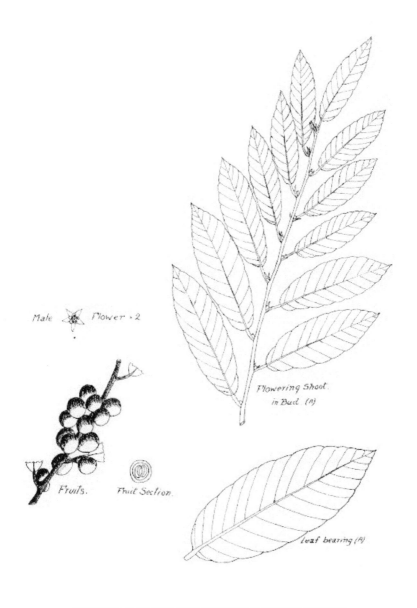

Male Flower × 2

Flowering Shoot in Bud (A)

Fruits. Fruit Section.

leaf bearing (B)

BURKEA AFRICANA Hook. f.—*Kolo, Kurdi, Bakin, makarfo, Namijin Kiriya, Kariye gatari*. LEGUMINOSAE.

A fairly large tree up to 50 feet in height with girths of 3-6 feet. It is a fine upstanding tree, frequently with a 20 feet bole, straight and clean, with ascending branches and high crown, spreading with age. It bears a resemblance to *Amblygonocarpus Schweinfurthii*, from which it can be distinguished by its larger and unequal-shaped leaflets. It is plentiful in high savannah forests and will extend, by means of small hollows or valleys as far north as 12½°. In flower or seed it is readily distinguished, and the leafless tree can be recognised by the thick, blunt end of the branches, often tufted with the dead stalks of last year's leaves.

The Bark is light grey with a bluish tint, and the scales are large and heavy, at times closely fitting, then falling in large sections, so that the trunk may appear shaggy. The branch bark, right up to the tips is very thick, with pronounced ridges and fissures and a light brown colour towards the ends. Often the leaf scars swell and form prominent knobs on the twigs. The slash is dark red and an oily resin exudes.

The Wood is reddish brown, the sapwood, light brown with an orange tinge. In transverse section the rings are very close, well-marked darker lines; the pores are very numerous, single and in festoons in well-developed soft tissue which shows up strongly in contrast with the hard tissue and nearly closes the pores. The rays are fine and evenly spaced, nearly straight, not visible to the naked eye owing to the amount of soft tissue. In vertical section the grain is close, the rings seen as bands of darker colour. The wood is very hard to saw and difficult to plane, but the planed surface is smooth and takes a high polish. It is liable to ring shake when seasoning, but is on the whole a sound, durable wood, very suitable for upright in constructional work. The weight is 72 lbs. a cubic foot.

The Leaves are some 18 inches long, at first erect, then drooping. They are bipinnate, with some four pairs of pinnae bearing 9-11 alternate leaflets, 1½-2 inches long and 1-1¼ inches broad, with unequal lobes and slightly cleft tip, bluish-green above, grey-green below, with a dull surface. The venation is prominent above and not beneath. The short leaflet stalk is curved.

The Flowers appear in March and April, mostly at the base of the new leaf shoot, a few only amongst the leaves higher up. They are in spikes or panicles up to 12 inches long, numerous, pendulous and conspicuous. The small white flowers are crowded and each consists of a minute calyx, 5 white petals which bend back in half, concealing the tips, 10 yellow-anthered stamens and a short blunt, curved pistil. The pentagonal shape of the corolla with its recurved petals and the erect stamens are noticeable. There is no scent.

The Fruits are pods, 2 inches long, 1 inch broad, thin, not flat, but tending to twist at the tip, forming a convex and concave side on the latter, of which the seed is clearly embossed. The colour is mahogany brown, veined, brittle and persistent. There is one brown, oval flat seed.

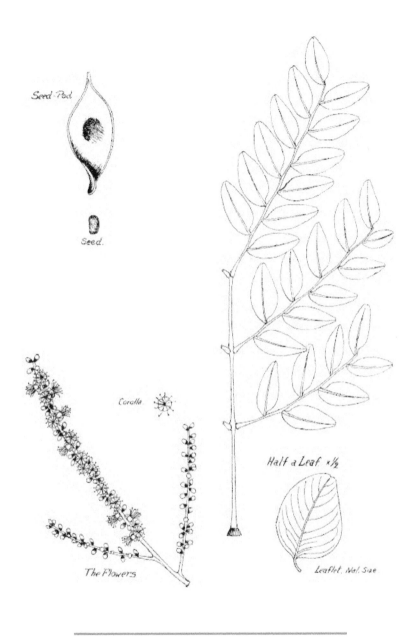

Seed Pod

Seed.

Corolla.

The Flowers

Half a Leaf × ½

Leaflet. Nat. Size

BUTYROSPERMUM PARKII Kotschy.—*Kadanya, Kadai. "Shea Butter Tree."* SAPOTACEAE.

This is one of the commonest species of the savannah forests and has a wide distribution. An average tree is about 30 feet high with a girth of 4-5 feet, though large specimens over 40 feet high with a girth of 10 feet are by no means infrequent. A short, stout bole and large, spreading limbs, gnarled and crooked, form a widely spreading crown of considerable density, which, from the drooping habit of the lower branches, reaches almost to the ground in many specimens. It is the type species of the tree savannah and in some parts of the country forms a large percentage of the forests.

The Bark is a distinguishing feature of the species and is dark grey, sometimes almost black, sometimes, particularly in the case of trees growing in barren situations, almost white. It has deep, vertical fissures and prominent, square scales of great thickness. The rough scaling extends to the quite small branches. A milky sap exudes from the crimson slash and from the leaf-stalks and twigs, when broken.

The Wood is a deep, dull red-brown with a purple tinge, the sapwood is pinkish. In transverse section the rings are indistinct lines and bands of dark and light tissue, the pores are small and in little rows between the fine rays, bands of soft tissue running in concentric lines and also connecting the chains of pores. The wood is extremely hard and heavy, difficult to saw and very hard to plane, the resulting finish being hard and taking a high polish. The grain picks up in part and the wood is liable to crack when dried. The weight is 80 lbs. a cubic foot.

The Leaves are strap-like, of an average length of 9-10 inches and a width of 2 inches; 3-4 inches of the length are occupied by the stalk and the veining is strong, these two features and the darker colour distinguishing this species from *Lophira alata* which it superficially resembles. The springing of the leaves from the end 2 inches of the twigs gives the idea of a rosette. The margin of the leaf is waved.

The Flowers are in round heads, 2-3 inches across, at the tips of the leafless twigs and appear from December onwards. Each flower is cream-white in colour, has 8 sepals in two rows, 8 petals, 8 stamens and a central, petal-like crown round the ovary, consisting of abortive stamens (staminodes). The flowers are sweet, with a rather nauseating perfume. As they die off, the young leaves shoot from the middle of the head. They are visited by bees.

The Fruits, which begin to form from January onwards are like green plums when ripe, and contain one or two shiny, chestnut-brown kernels, with a large scar, or hilum, having the appearance of a split seed coat. They are about 2 inches long and 1 inch wide, the thin seed coat enclosing a firm, white flesh, from which the oil is extracted.

Uses.—The oil is extracted and the butter made by the natives by a preliminary boiling in water, followed by repeated pounding and stirring in cold water, the oil which rises to the surface being skimmed off and placed to harden in calabashes. It is eaten as such, burnt as an illuminant and used as a base for certain medicines.

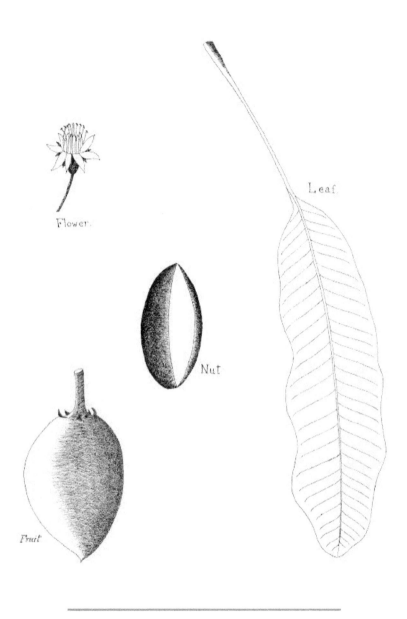

Flower.

Leaf.

Nut

Fruit

CASSIA AREREH Del.—*Malga, Gamma fada.* LEGUMINOSAE.

A small, erect tree from 15-20 feet in height, about a foot in girth, with one or more stems and a high, drooping crown of delicate pinnate leaves. It resembles *Cassia Sieberiana* only in the shape of the flowers, leaves and pods, and the outstanding differences are the erect form, the perfume and the longitudinally splitting pods. It grows in good soils and not in the barren places where *C. Sieberiana* abounds.

The Bark is a dull grey, with wavy but not prominent ridges of crisp bark, often rather shaggy when the scales fall in long pieces. The slash is pale brown.

The Leaves are pinnate, about 10-12 inches long with 6-7 pairs of leaflets some 2 inches long and narrowing in proportion to the length towards the top of the leaf. The main rib is very slender and is produced beyond the top pair of leaflets in the form of a slender curve, which frequently replaces one of the final pair of leaflets. The leaflets are smooth, clearly veined and purplish when young.

The Flowers are in racemes some 6-10 inches long, crowded, especially at the tip, with the flowers whose stalks grow to 3 inches in length. The three bracts, one long and a pair shorter are very conspicuous, especially at the crowded tip of the flower stalk where the buds are purplish. The flowers are typical of the genus, with 5 curled sepals, 5 irregular oval petals, 3 long, 4 medium and 3 short stamens, and a long, curved pistil swollen in the middle. The flowers are highly perfumed and appear in February in large masses amongst the new leaves.

The Fruits are pods, 12-20 inches long and an inch thick, cylindrical with irregular constrictions, purplish black in colour, shiny and hard. They persist on the tree for a long time and finally split down one side to release the light brown seeds. The inside of the pod has a yellowish, dry, mealy pulp.

Uses.—The pulp of the pod is used as a laxative.

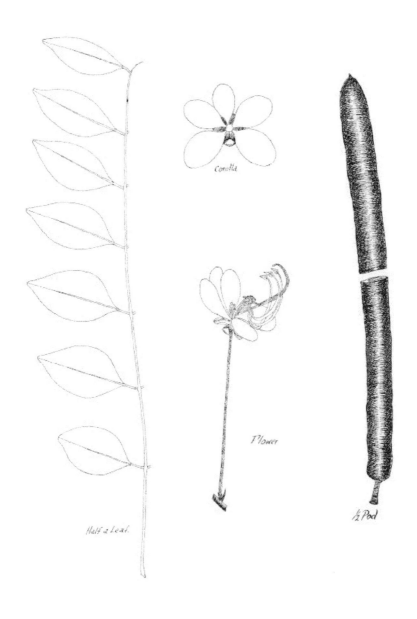

Corolla

Flower

½ Pod

Half a Leaf.

CASSIA GORATENSIS Fres.—*Rumfu. Runhu.* LEGUMINOSAE.

This is a small and common tree whose only value, apart from medicinal use by the natives, is ornamental. It is found in clumps as a small tree about 10 feet high, but also singly growing to a height of 20 and even 30 feet. Well-grown examples have a 10-12 feet bole some 3 feet in girth and a spherical crown. In its young state it is apt to be confused with *Cassia occidentalis*, "rai dore," which latter is a shrubby herb with similar flowers and leaves.

The Bark is grey, sometimes almost white, with wide fissures and large scales. Old trees may have very dark and rough bark. That of the stout tapering twigs is a light brown with a powdery surface which rubs off in the hand. The slash is dull brown.

The Leaves are pinnate, 10-12 inches long with about eight pairs of oval leaflets which are bright green and smooth on the upper surface; grey, with the venation raised on the under surface. They are soft in texture and very often blotched or spotted.

The Flowers are most conspicuous in large, round bunches at the twig ends. They are bright yellow and about 2 inches across, with 5 petals, of which the lower three are much the same size and shape, while the other two are larger, the one spoon-shaped, the other serrate-edged and with marked veining. The sepals are also yellow and saucer-shaped, two of them larger than the other three. Of the 10 stamens, 3 are tall with large anthers, 4 are short with large anthers and 3 are quite small with round, flat anthers. They are grouped accordingly, from back to front as is typical of the genus. They appear in September.

The Fruits are thin, brown, jointed pods about 4 inches long, straight or curved, hanging in bunches. They contain 15 or 20 small, round, flattish grey seeds and ripen in February.

Uses.—The pods and leaves, boiled in water, are used for washing and purification after child-birth. The same infusion is taken internally as a cure for fever.

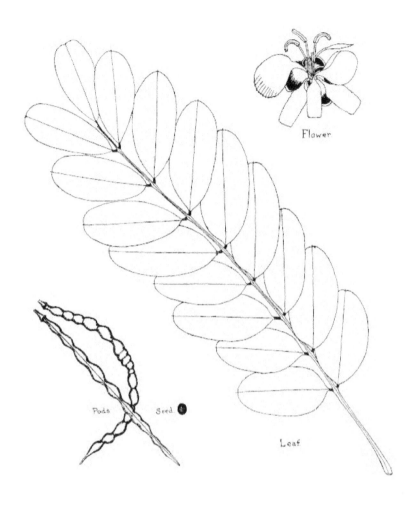

Flower

Pods   Seed

Leaf

CASSIA SIEBERIANA DC.—*Malga, Gamma fada.* LEGUMINOSAE.

A very common tree with a wide distribution, growing on any kind of soil from rich loam to the driest and most barren sand or clay. It is commonest on wastes swept bare and scoured by rain or in little stony gullies. The form varies with the locality, tall, slender trees growing on good soils and wide-spreading, many stemmed shrubs of great size, or small stunted trees on poor soils. The full-grown tree is some 30 feet high, branching from ground level and forming a hemispherical growth with foliage to the ground.

The Bark of young trees is black, smooth and covered with lenticels that are rust-coloured on the smaller branches. That of old trees is dull black and rough with large heavy scales several inches long, that give the stem a shaggy appearance when they are falling. The slash is a yellow ochre colour.

The Wood is pale red, darkening to a light mahogany brown after exposure. The sapwood is white. In transverse section the rings are ill-defined, though the hard and soft tissue is well marked and the colour is much deeper in this section. The pores are numerous in long chains and festoons; the rays, invisible to the unaided eye, are fine and nearly straight. The wood is very hard and heavy and difficult to saw and plane, the finished surface taking a bright polish. Weight 70 lbs. a cubic foot.

The Leaves are some 12 inches long, pinnate with 6-9 pairs of leaflets opposite or nearly opposite. These are varied in shape, the basal pair almost as broad as long and the top pair almost twice as long as broad, with graduations between, 1½-2 inches long and 1-1¼ inches wide. They are bluish-green with a brilliant sheen, paler beneath with the mid-rib prominent. Young leaves are often purplish. The petioles are ¼ inch long.

The Flowers are in 12-18 inches long, pendulous racemes, in masses with the appearance of Laburnum, from February to May. Each flower is 1½-2 inches across on a 1½-2 inch stalk, and has 5 small, irregular sized pointed sepals with a darker green central line, 5 large yellow petals, varying in shape, and 10 stamens, 3 long and large-anthered, 4 shorter and large-anthered, and 3 small and round-anthered. The pistil is large, smooth and green. The petals enlarge and pale after the stamens have fallen from the fertilised flower. There is no scent.

The Fruits are long, black pods, up to 2 feet in length and a little over ½ inch broad, cylindrical, straight or slightly curved, more or less jointed, but never deeply, and divided across the length of the pod by fine, transparent, brittle membranes forming cells up to a hundred in number containing oval, shiny, light brown, hard, ½ inch long seeds, one in each cell, rattling loose in the pod. The pods fall entire or break across and rot on the ground. They are mostly riddled with holes by a small grub. The dry, yellowish pulp has an unpleasant smell.

Uses.—The yellow, mealy pulp is used in the preparation of a laxative.

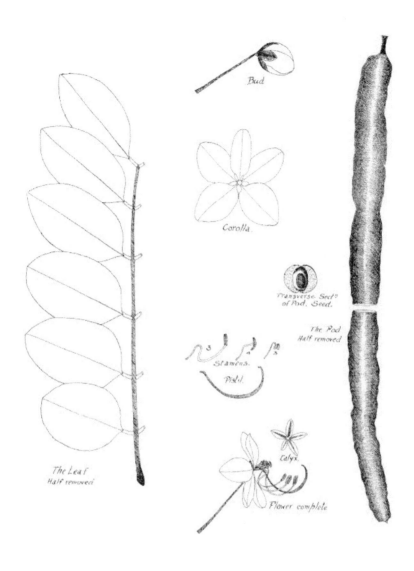

Bud

Corolla

Transverse Sectⁿ of Pod. Seed.

The Pod Half removed

Stamens.

Pistil.

Calyx.

The Leaf Half removed

Flower complete

CELTIS INTEGRIFOLIA Lam.—*Zuwo, Dukki*. ULMACEAE.

A large tree which, when full grown attains a height of over 60 feet and a girth of 12-15 feet. It is not commonly distributed all over the country but is very common locally, *e.g.*, in N.E. Sokoto and in N.W. Bauchi, in both of which places a large number of big trees are found over a small area of country. It may have a clean bole up to 20 feet or more in length with, in old trees, root flanges. The species is so frequently lopped for fodder that it is rare to find a naturally formed tree except in uninhabited forest where it produces a very large crown rounded at the top and, as a rule, branching low down. The limbs are large and bent; the twigs slender, horizontally spreading or drooping. The foliage is dense but not heavy. There are often thickets of adventitious shoots on the trunk.

The Bark is a light, slightly bluish-tinted grey, fairly smooth, with large, thin, very hard, rounded scales which come away from the lower end, remaining attached to the tree at the upper end and giving the rather curious look of sliding down the tree. Under the scales the bark is light brown. The slash is mottled, dark brown crumbling, white fibrous.

The Wood is light yellow in colour, soft, fairly close-grained, easy to work with all tools, planes to a dull, smooth finish, cracks only slightly when seasoning, is not durable under exposure, being inclined to bluish discolorations and weighs only 35 lbs. per cubic foot. The native does not use it, but it seems a useful soft timber. The pores are very numerous, of different sizes, some quite large, open and in festoons. The rays are just visible with the unaided eye and the rings plainly seen as close fine light lines. The rays are much waved.

The Leaves are variable in shape and size, some toothed and some plain along the margin. They are from 2-4 inches long and from 1-2 inches wide with pointed tip and ¼ inch stalk. The serrate leaves are found in numbers on the adventitious twigs, the entire leaves generally on the long, slender twigs which bear the flowers. Old leaves are very much darker in colour and rougher to the touch than younger leaves. Both surfaces are covered with very short, stiff hairs and are rasping to the touch. There are three main veins and the leaf is not symmetrical in shape. They turn their upper surface to the sky and are arranged alternately on the twig.

The Flowers which appear from December onwards are in minute, branched clusters in the leaf axils. Each flower is green in colour and has 5 sepals, 5 stamens with yellow anthers, and a twice forked stigma which enlarges as the fruit grows and then withers. The remains of the sepals and the stamens are found at the base of the fully grown fruit whilst it is green.

The Fruit is a plum, single-stoned and fleshy. If cut through before ripe, the seed can be seen growing in the top under the stigma and it expands downwards from this point, only partly filling the cavity when ripe. The ripe fruit is light brown and ribbed, about ½ inch long, and the stone is white, very hard and marked all over with a raised network.

Uses.—The leaves are a valuable fodder for cattle and made into soup by the natives.

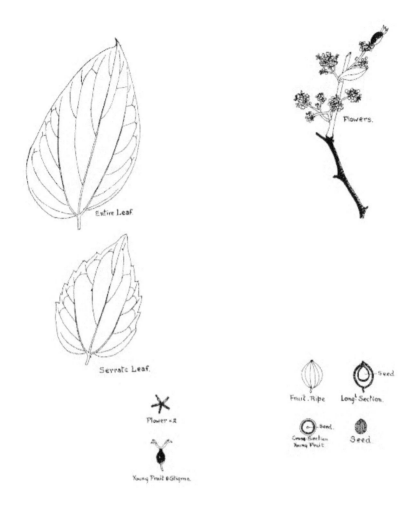

Entire Leaf

Serrate Leaf.

Flowers.

Flower × 2

Young Fruit & Stigma.

Fruit. Ripe   Long.¹ Section.   Seed

Cross Section.   Seed.
Young Fruit

Seed

COMBRETUM ABBREVIATUM Engl.—*Kariya.* COMBRETACEAE.

This very beautiful species is a climbing or spreading shrub with a multitude of stems forming a tangled thicket with occasional stems extending into the air some 30 feet in length. It can be distinguished by its masses of brilliant red flowers and by the great variety of its leaves and the pale colour of its fruits. It is not common, so mention is made of two places where it may be seen, namely, in a "kurmi" at Rahamma (Zaria Province) and at Dan Tudu, near the G. Mainu Reserve on the River Rima (Sokoto Province).

The Leaves vary very considerably from 3-7 inches long and 2-4 inches broad. They may be almost as broad as long or twice as long as broad. On the new shoots there are short branches, modified to the form of blunt spines at the end of which are single leaves. When these drop off the thorn hardens but is never very sharp. These leaves are arranged spirally round the stem, unlike those of the other branches without thorns, which are inclined to one plane. The leaves are dark green and smooth and the venation shows up almost white in contrast. The venation is prominent on the under surface.

The Flowers are in enormous panicles several feet in length at the end of the shoots. The clusters vary in size at close intervals all the way along the shoot and the flowers themselves all stand up in the same vertical position, forming a great flat mass of brilliant red bloom. They appear in February and each consists of a light green 4-lobed calyx, 4 small red petals meeting at the tips and overlapping. From the tips of the petals 8 stamens emerge, 4 long, 4 short, bright red, as is the pistil.

The Fruits are 4-winged, ¾-1 inch in diameter, a delicate green when new and a light brown when ripe. The single seed is long and 4-angled. They are ripe in March.

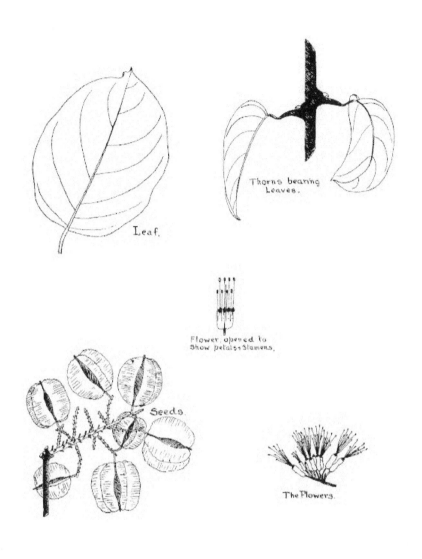

Leaf.

Thorns bearing Leaves.

Flower, opened to show petals, Stamens.

Seeds.

The Flowers.

COMBRETUM HYPOPILINUM Diels.—*Jan Taramniya, Jan Ganyi.*
COMBRETACEAE.

A small tree, rarely exceeding 20 feet in height, except in good soils, found very commonly in the drier country of the extreme north or in open savannah where the *Terminalia* species abound. It is not unlike *C. lecananthum*, but the seeds are larger and the leaf broader. Its form is typical of the small, wide-spreading trees of the Bush savannah, one or more stems and a number of spreading branches forming an open crown. Occasionally, though rarely, the species exceeds 20 feet and has an 8 feet bole. The tough, deep red fruits are distinctive.

The Bark is grey or brown, smooth in young stems and later showing long, vertical, narrow ridges and stringy fissures. The slash is red-brown.

The Leaves are variable in size, up to 5 inches long and 2½ inches broad, oval with a pointed tip and a stout ¾ inch long stalk. The venation is prominent beneath and flush on the upper surface, the mid-rib very prominent and light coloured. The upper surface is dark green and smooth, the underneath grey-green. They are tough in texture.

The Flowers which appear from December to February are in axillary spikes up to 3 inches in length crowded with cream-coloured, slightly scented flowers, each of which has a 4-pointed calyx, 4 minute petals, 8 stamens and a short pistil rising from a calyx full of white hairs.

The Fruits are 4-winged, 1-1¼ inches in diameter and length, conspicuous by their size, tough texture, and the rich, reddish-purple colour softened by a grey bloom. They are very numerous, crowded, heavy and hard.

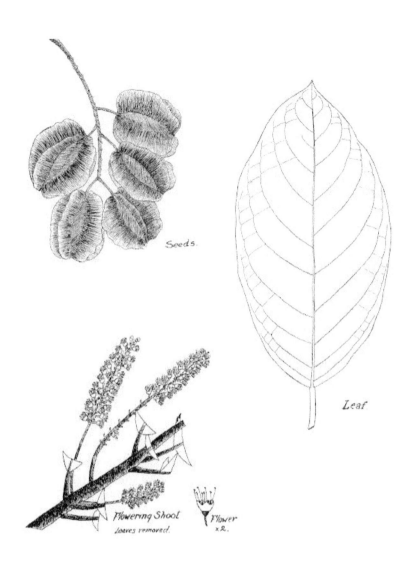

Seeds.

Leaf

Flowering Shoot
leaves removed.

Flower
x 2.

COMBRETUM LECANANTHUM Engl. & Diels.—*Chiriri, Dugera.*
COMBRETACEAE.

A very common tree of the more open, dry savannahs and occurring on hard and barren or dry and sandy soils, often in such quantities as to form almost pure forests. It occurs mixed with the shrubby *Combretum* species or by itself in belts and patches on slightly better soil than is required to support the latter. It is about 20-25 feet high but a full-grown example will reach 40 feet with a 5 feet bole girth. The stem is rarely straight, often leaning or twisted and branching low down. The crown is either of erect branches and twigs, or the latter may as often as not droop and hang vertically downwards.

The Bark is smooth, grey-brown or rust-red in colour, with hardly any scales, but of a rather fibrous appearance. From it, in the hot season, large quantities of gum exude, white, yellow or red-brown, poor in quality, not brittle, but rubber-like, melting in moist heat and drying only in the dry season. The gum forms strange shapes, from nodules to long slender spirals. The bark is very thin. The slash is crimson, with yellow centre.

The Wood is a dirty white or cream colour. In transverse section it is much darker and the pores are very small and in little wavy lines and festoons, the rays exceedingly fine and close light lines. In the plank the pores are fine dark lines. In the green state the wood is so tough as to blunt axes but when it dries it loses its strength and is often completely destroyed by borer beetle. An unsatisfactory wood of no value. Weight 56 lbs. a cubic foot. The name "kariye gatari," or "break axe" is given to it.

The Leaves vary largely in shape and in proportion of length to width. They average 4-5 inches long and 2-2½ inches wide, tapering both ends, pointed, shining above, paler beneath, the venation simple and the lateral nerves inclined far forward. They are in pairs and tend to lie in the same plane.

The Flowers are in branched spikes, cream-coloured and sweet-scented. They appear in November and may flower up to July. Each has the 4-pointed calyx, 4 minute petals, 8 stamens and short pistil common in this genus. There are bright orange glands inside the calyx.

The Fruits are 4-winged seeds, about ¾ inch long. They ripen from green through red to brown, and hang in loose clusters amongst the leaves. The stalks of the flower spikes elongate considerably with the seed growth and intermingle, forming a tangle of seed and stalk.

Uses.—There are none other than that the gum is chewed by the natives.

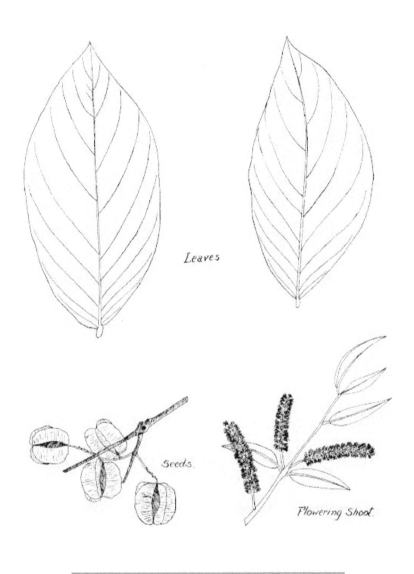

Leaves

Seeds

Flowering Shoot

COMBRETUM LEONENSE Engl. & Diels.—*Wuyan damo*.
COMBRETACEAE.

This species is found distributed fairly evenly over large areas of both Tree savannah and the drier, more open forests, and is common in rocky country. It has a tendency to grow longer, straighter boles than other species of *Combretum* and is commonly met with 20-30 feet high, or even more, with girths of 3-4 feet. If surrounded by other trees, the stem is exceptionally straight and readily distinguished by its dark colour and the small, square, regular scales. The crown, with its large, dark, drooping leaves, is dense, and big trees afford good shade. The branches droop over and the long, slender twigs hang straight down.

The Bark, a ready means of identification, is dark grey, sometimes almost black, and is covered with small, prominent, square, even-sized scales from which the tree gets its Hausa name, owing to its resemblance to the skin of the large lizard. It yields a gum. The slash is yellow.

The Wood is yellow or greenish-yellow. In transverse section the pores are small, in groups and festoons, the rings indistinct and the rays so fine as to be invisible. In the planks the pores are long, wavy and dark, the grain being crooked and fibrous in appearance. It is a very hard, tough, heavy wood, most unsatisfactory to saw and impossible to plane. The strength and durability are great. The weight is 64 lbs. a cubic foot.

The Leaves are larger than those of other species and are dark green, with a softly hairy surface and a soft texture. They are about 7 inches long and 2½-3 inches wide, with a long, slender tip, a taper at both ends and a very short stalk. They are pendulous on the long, slender twigs.

The Flowers are in clusters of 1½ inch spikes and are yellow with a sweet odour of musk. There are 10 yellow stamens round a square calyx and the pistil is surrounded densely by hairs on the receptacle. The flowers appear in March.

The Fruits are the typical winged seed of the genus and are smaller than those of most other species, a light brown in colour, less than 1 inch long and very persistent.

Uses.—The stems of small trees make excellent posts for use in the ground, owing to their great durability.

A concoction from the bark is used medicinally as an astringent.

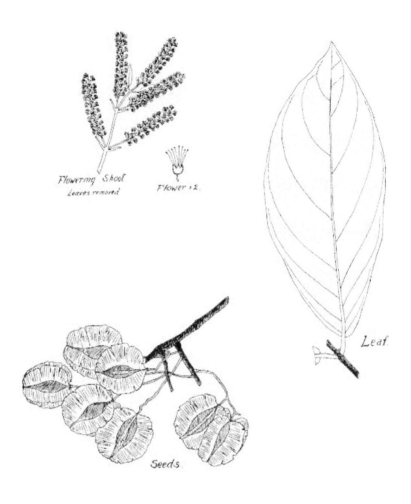

Flowering Shoot
Leaves removed

Flower × 2.

Leaf

Seeds.

COMBRETUM MICRANTHUM Don.—*Fara Geza*.
COMBRETACEAE.

This is one of the commonest species in the north, and covers many square miles of the driest and most barren of rocky wastes where little else will grow. The stout root stocks and tough, wiry stems survive fire, drought and excessive heat. It grows from 3-10 feet high, pure or mixed with other *Combretum* species or with *Guiera senegalensis*. It is distinguished by its generally smaller leaf, smaller fruits (seeds), lighter green foliage and the masses of highly perfumed flowers.

The Leaves are light green and shining, in pairs. The tip is pointed and the margin sinuous. They decrease in size from the base of the twig upwards.

The venation is raised on both surfaces. They are 2-3 inches long and 1-2 inches broad.

The Flowers are in dense masses of spikes in the leaf axils from May onwards and fill the air with perfume. Each has a 4-pointed calyx, 4 small petals, 8 stamens and a short pistil. There is a ring of orange-coloured hairs round the pistil.

The Fruits are 4-winged, ½ inch in diameter and length, brown when ripe, in dense clusters on brittle stalks. They are persistent right through to the following rainy season.

Uses.—The bark is stripped for fibre and the stems used entire for binding the rafters of grass roofed houses or split for basket making. The stems are so tough that an axe or matchet will as often as not fail to cut through, but split the stem.

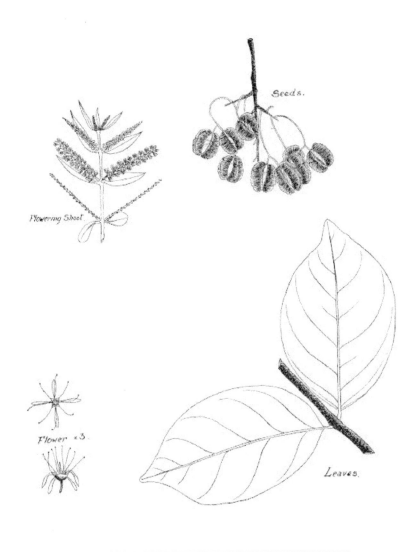

Flowering Shoot

Seeds.

Flower ×3.

Leaves.

COMBRETUM VERTICILLATUM Engl. & Diels.—*Farin Taramniya.*
COMBRETACEAE.

This is one of the commonest of the larger *Combretum* species, especially in the drier savannahs. It is particularly common in Sokoto Province and can be distinguished by the grey, dusty-looking leaves. There are two common forms of growth. The more common is that with a bole from 5-10 feet long, or several stems from ground level, and a round straggling open crown with the twigs drooping low down. The other is the tall, erect stem up to 40 feet in height, branching to within a few feet of the ground but having a cylindrical crown composed of erect limbs terminating in long slender branches of great length extending high into the air or dividing into numerous drooping flaccid twigs. The heavy leaves sway and weight down the weak twigs which can be bent to a circle without breaking. The species grows in the poorest sandy soil.

The Bark is pale grey or brown with very small, crisp scales scattered all over it, there being wide spaces between the scales, especially in the spring. The scales are dotted with small rust-coloured lenticels. The bark is sometimes cream-coloured. The bole is rarely cylindrical, but most often columnar and having rounded root flanges. The slash is red-brown.

The Wood is yellow or greenish-yellow. In transverse section the pores are small, single or in small groups, and the rings are very faint, the rays hardly visible, except as small brown bands in radial section. In the plank the pores are brown and open and the grain irregular, with black lines, flecks and brown patches. An unsatisfactory wood, cross-grained and tough, but strong, hard and fairly sound. Difficult to saw and plane. Weight 55 lbs. a cubic foot.

The Leaves vary much in size from 3 inches long and 2 inches wide to 5 inches long and 2½ inches wide. The base may be slightly cordate or narrowing and the tip is generally abruptly pointed. The margin is wavy, the leaf rarely flat and the foliage is very subject to attacks by insects and is much contorted, indented and disfigured. The upper surface has a dusty grey bloom and the lower surface is grey. The venation is raised on both sides of the leaf. The whole is soft and velvety to the touch. There is a ¼ inch stalk.

The Flowers are yellowish and in racemes, the spikes 1-2 inches long. They first appear when the tree is leafless from December onwards and are scented. Each consists of a 4-lobed calyx, 4 small white, recurved petals and 8 stamens opposite sepals and petals.

The Fruits are 4-winged, 1-1¼ inches long, all shades of red till they are ripe, when they turn brown. The seed is long and 4-angled.

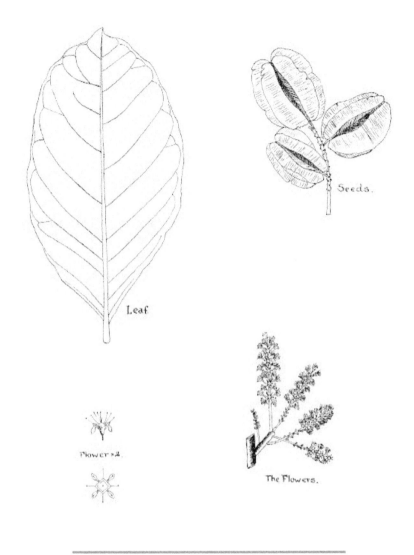

Leaf

Seeds.

Flower ×2.

The Flowers.

CORDIA ABYSSINICA R. Br.—*Aliliba*. BORAGINACEAE.

A small tree up to 25 feet high, either shrub-like in habit or with a short bole and round crown. The distinguishing features are the large heart-shaped leaves, large heads of papery flowers and the clusters of yellow fruits. As a shrub the branches spring from ground level and form a large compact bush, and as a tree the bole may be from 6-10 feet long and some 2-3 feet in girth.

The Bark is light grey and smooth with small lenticels. Only on the larger trees do scales form and the bark splits longitudinally and has a fibrous appearance. The slash is white.

The Leaves vary greatly in size from the quite small ones at the twig ends to the large basal leaves a foot long and 9 inches wide. They are heart-shaped, with stout stalks, the mid-rib curved back and the leaf tending to fold up along its length. Both surfaces are rough with short hairs. The upper surface is dark green, the lower paler. The venation is bold, the mid-rib not straight, but changing direction at each lateral nerve. The lateral nerves are connected by parallel veins. The leaf is tough in texture.

The Flowers which appear in October, are in large branched heads, a mass of buds prolonging the flowering period. The flower blooms for one day and then dries crisp, retaining its shape more or less when dry. Each flower is an inch across, funnel-shaped, white, with veined and crinkled corolla with 5 indistinct lobes. There are 5 black-anthered stamens and a twice-forked style with 4 stigmas. The 5-toothed calyx is a short tube and is heavily ribbed. The flowers have no stalks and there is a slight perfume.

The Fruits are in clusters; round juicy drupes with a hard stone and yellow translucent flesh, edible and sweet. They are ½ inch in diameter.

Uses.—The fruits are eaten, either fresh or mixed with honey to make a sweetmeat "alewa."

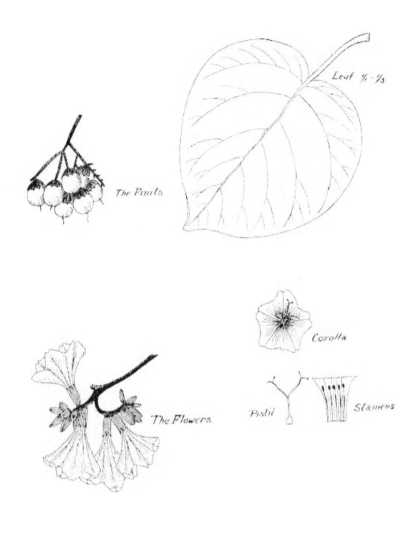

The Fruits

Leaf ⅓-⅔

Corolla

The Flowers

Pistil

Stamens

CRATAEVA ADANSONII DC.—*Gude.* CAPPARIDACEAE.

A medium sized tree found growing locally in considerable quantities in "fadammas." It attains a height of about 30 feet with a girth of 4-5 feet. Owing to the fact that the leaves are cropped as a human foodstuff, few trees retain their natural form and the general appearance is that of a pollard willow, with a short, stout stem and a number of erect slender shoots, inclined to bend down. Owing to its being liable to submersion for certain periods, the stem is frequently in a semi-recumbent position and from it spring a number of shoots forming a false crown. It is a very handsome tree in full flower and the delicate foliage is attractive.

The Bark is a light brown with a few crisp scales here and there. That of the branches is smoke-colour, densely covered with light brown lenticels. That of the new twigs is pale brown.

The Wood is very soft and of a rich yellow colour. It has a strong not unpleasant odour. It is of no practical use.

The Leaves are trifoliate with a stalk some 3 inches long and the leaflets from 3-4 inches long and an inch wide. They are narrow at both ends, light green and smooth, tender in consistency. They are in whorl-like bunches at the tips of the twigs.

The Flowers are very handsome and appear in February at the tips of the twigs in clusters of 10-20. Each has 4 small pale green sepals, 4 large leaf-like white petals grouped round one-half of the calyx and some 18 inch-long mauve stamens with mauve anthers. The pistil, which appears prominent in length after fertilisation is some 1½ inches long, with a knob at the end.

The Fruit is very like that of *Strychnos* (kokiya) in appearance, a pale brown sphere from 1½-3 inches in diameter on a woody stalk about 2½ inches long. The inside contains a number of small dark brown seeds of quaint "curled-up" shape embedded in a white, mealy flesh. The "rind" of the fruit is thin and crisp. The fruits are ripe about November onwards. They are eaten by birds who pick out the contents and leave the empty "shell" on the tree.

Uses.—The leaves are considerably eaten by the natives, and in times of famine are taken great distances to the markets.

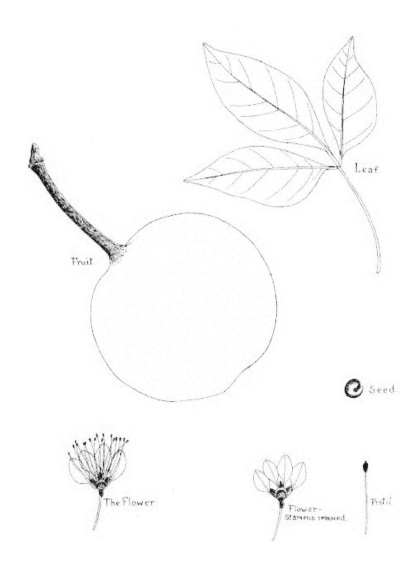

Leaf

Fruit

Seed

The Flower

Flower - Stamens removed

Pistil

CROSSOPTERYX KOTSCHYANA Fenzl.—*Kasfiya. Kashin awaki.*
RUBIACEAE.

This fairly common tree occurs more plentifully in localities which are flooded during the rains and baked hard and burnt bare in the dry season. Thus it may take the place of *Mitragyne africana* (giyeya) in wet-season fadammas where the grass is rank in growth. It is not, however, confined, like the above, to such situations, but abounds in open savannah and rocky places. It is most readily identified by its fruits, q.v. and also by its habit of sending up a number of stems from a common stock, these stems growing erect and close together and showing an irregular cross section, oval, grooved or flattened in shape by reason of this close proximity to each other. In form the tree is cylindrical with a rounded or pointed top and sparse foliage. The foliage extends almost to the ground. A height of 30 feet and 2-3 feet girth is usual.

The Bark is peculiar in structure and is light grey or brownish, with small grey scales which will crumble in the hand and appear to consist of numerous lenticels. It is smooth and the scales are not prominent. The slash is salmon pink and crumbling.

The Wood is a brown pink, not unlike Pear wood. In transverse section the pores and rays are scarcely visible, so close is the texture. The rings show as faint light lines, close together, and the minute pores are scattered about between the exceedingly fine and numerous rays. The grain shows as slight banded variations in depth of colour. It is a sound, hard wood which seasons well, saws and planes easily to a hard, smooth finish, taking a good polish. The small sizes are unfortunate. Weight 57 lbs. a cubic foot.

The Leaves are about 3½ inches long and 2 inches broad, alternate and inclined to assume one plane. They are pale green and downy on the upper surface, roughly and intricately veined beneath with light brown velvety hairs on the veins.

The Flowers are in close bunches with a tendency to droop from their weight; each flower consisting of a small 5-lobed calyx, a long tubular corolla with 5 petals, white tinged with pink, and a long, clubbed pistil surrounded by 5 small, protruding stamens. They are sweet-scented and appear in February.

The Fruits are easily recognised, forming bunches similar to the heads of flowers. Each is a small blackish capsule with a black tip, and it splits into two sections across the two seeds which are the shape of half the capsule and are black. They are very persistent and remain on the tree for a long time after they have shed their seeds.

Uses.—The stems are cut for poles (gofas).

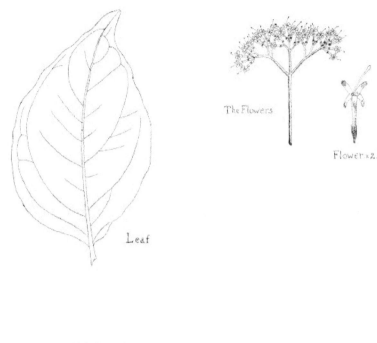

The Flowers

Flower ×2.

Leaf

Fruit Section ×2

Capsule ×2½

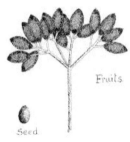

Fruits.

Seed

CROTON AMABILIS Müll. Arg.—*Koriba.* EUPHORBIACEAE.

A small, erect tree from 10-25 feet high which will grow in the extreme north in the driest and most barren of soils. It flourishes round the parallel 14° north and extends right down to the extreme south also. It is often a stunted, shrub-like tree growing in quantities on bare, hard, stony soils, but in more open sandy localities will grow a 15 feet bole with an open light crown composed of erect, slender branches. Old trees spread wide and give a little shade. It can be at once distinguished by the silvery glister of its leaf on the under side and by the spikes of closed flowers.

The Bark is grey or light brown with small, close-fitting, even-sized polygonal or rectangular scales.

The Wood is pale yellow, discoloured grey in patches. The rings are fine, close, light lines, the pores are very fine, numerous and evenly distributed, connected by very fine lines of soft tissue, narrower than the pores. The grain is fine and the wood is hard, difficult to saw, not hard to plane and finishes with a hard, smooth surface which will polish. The weight is 64 lbs. a cubic foot.

The Leaves are long, narrow and pointed, 4 inches long and 1½ inches broad with a 1½ inch stalk. The upper surface is a dark shining green, the lower glistening silver with minute rust spots all over the surface. The mid-rib is sunk on the upper and very prominent on the under surface. The lateral nerves are numerous and parallel. The stalk and mid-rib are covered with soft brown hairs. The young leaf is bronze or red-coloured. The leaves droop and tend to fold up along the mid-rib.

The Flowers are of two sexes, each on a different tree (dioecious). They are in racemes, resembling long spikes. The male are the longer, up to 4 inches, the female up to 2 inches. The male flower has 5 sepals, 5 petals and some 15 stamens all packed up in the bracts, on short stalks. The females have similar flowers with 3-part styles each with a bifid stigma. The flower parts are hardly visible to the naked eye so closely folded are the bracts.

The Fruit is a capsule, 3-lobed, one seed in each lobe. They ripen about November onwards, but may be found for many months in the year.

Fruit from above

Fruit

Section of Fruit.

Flowering Shoot

CUSSONIA NIGERICA Hutch.—*Gwabsa, Takandar giwa.*
ARALIACEAE.

This quaint looking species is locally very common, more especially on rocky hills where, in its leafless condition, it resembles some cactus growth. It is not, as a rule, above 20 feet in height, but may reach over 30 feet with a girth of 4-5 feet. The girth is large in proportion to the height and there is practically always a clean bole, sometimes divided at the base into two stems, with an umbrella-like crown, flat-topped. The form of the tree is best seen in its leafless state, when it has the appearance of innumerable stags' horns, the branches being very thick and blunted at the ends. In leaf, its more or less grotesque appearance is disguised by the large and handsome foliage.

The Bark is very rough and vertically fissured, light grey with long, corky ridges and regular sized scales. A clear gum exudes from a cut. Annual fires are harmless but cause a large amount of charcoal on the outside so that the stems are nearly always blackened. The slash is pale brown.

The Wood is a dirty grey colour, very soft, brittle and easy to work, is not durable, rots readily and weighs only 23 lbs. a cubic foot. In transverse section the rings are dark lines at broad intervals and the rays are clearly visible to the unaided eye. The pores are open and distributed in small rows, singles and groups between the rays; a useless wood.

The Leaves are digitate with about 8 digits, in length from some 10-20 inches in the one leaf. They spring from the tips of the blunt twigs in an erect bunch, at first almost purple in colour. Each has a 2 feet long stalk and the edges are serrate. They are pale green in colour.

The Flowers appear in February and are on stout spikes of weird shape and appearance, which grow well over a foot long and are about one inch thick. The flower spike shows flowers and fruits in all stages of maturity from the bud to the seed and may be straight or twisted and tangled up with the others on the twig. Each flower is pale green with 5 petals, between ½ and ¾ inch across, with 5 stamens and a bifid stigma on a stout ovary.

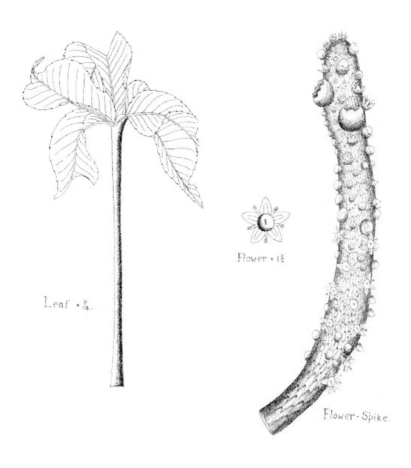

Leaf × ¼.

Flower × 1½

Flower-Spike.

DETARIUM SENEGALENSE Gmel.—*Taura*. LEGUMINOSAE.

In the north this species is generally quite a small tree from 15-20 feet high with a few erect branches forming a small flat-topped crown. Often, however, quite large trees can be seen, even in the driest country, and further south large specimens up to 40 feet high with girths over 6 feet are not uncommon. In the southern provinces, the same tree will grow to a height of 80 feet with a 12 feet girth. The larger portion of the height is almost always the crown, which is spreading and dense, giving good shade. The tree is readily distinguished by its bluish bark and round, compressed fruits. The small trees show enlarged tips to the branches from which the leaves spring. In places it is so common as to form almost pure forest over small areas.

The Bark is bluish-grey, with large polygonal scales. On young trees there are yellowish patches where the scales have fallen and the bark of the smaller branches is ochrous and powdery. The slash is pale crimson.

The Wood is dark brown or red-brown. In transverse section the rings are slightly darker lines, the pores are small, clear and open, mostly single, some in groups or nests of 2-4, rather unevenly and widely scattered. The rays are visible, not all continuous, some broader than others, evenly spaced with room for the pores between, showing in radial section as brown bands, so conspicuous on the sapwood as to colour it. In vertical section the pores are light coloured in the heartwood and dark in the sapwood, in which the resinous pore contents darken the wood in rings, showing as lines in the vertical section. The wood is tough and hard and not easy to work, though it has a slight sheen and takes a polish. The weight is 55 lbs. a cubic foot.

The Leaves are pinnate, 6 or 7 inches long with 6-10 alternate or opposite elliptical leaflets some 3 inches long and 1½ inches broad. Those nearer the top are more oval, those at the base rounder. The tip has a slight cleft. The leaf-stalks are very short and stout and covered with dusty brown hairs. The surface is waxy, there may be a few hairs, and bluish-green. The texture is rather leathery.

The Flowers, which appear in May, are found in masses all over the tree. They grow on short, branched stalks in dense clusters. There are no petals, their place being taken by the white petal-like sepals, 4 in number. There are 8 short, curved stamens with yellow anthers and a short pistil.

The Fruit is a drupe about 1½ inches across and flattened to about ¾ inch. It has a brown outer skin, a greenish, mealy flesh full of fibres which when the flesh decays remain round the kernel. This flesh is sweet and edible, but not very palatable. On small trees clusters of abortive fruits are found; little round, brown-green, soft growths which do not mature.

Uses.—The flesh is used by the natives in the manufacture of the sweetmeat "madi."

Where it is large enough the timber is used for mortars, but the size is, as a rule, against this use.

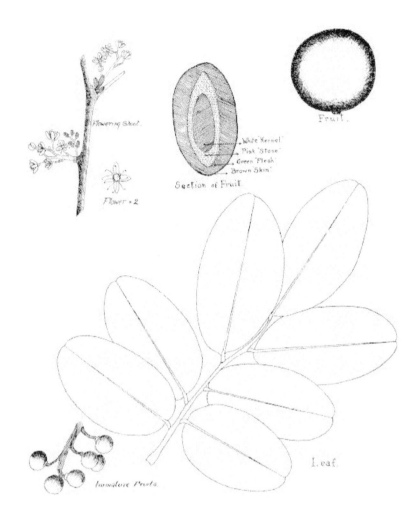

Flowering Shoot.

Flower × 2

White Kernel
Pink Stone
Green Flesh
Brown Skin

Section of Fruit

Fruit.

Immature Fruits.

Leaf.

DICHROSTACHYS NUTANS Br.—*Dundu*. LEGUMINOSAE.

A very common, acacia-like shrub which occurs in dense thickets in open situations. It averages about 10 feet high, though it will attain nearly 30 feet, branches near the ground and is most readily distinguished both by its flowers and pods. It is of no use in the composition of the forest, but is quite ornamental in flower.

The Thorns are modified branches, each twig ending in a thorn which bears leaves, and is almost black in colour.

The Leaves are bipinnate, up to 4 inches long with 10-15 opposite pinnae, each bearing 15-20 pairs of dark, dull-green leaflets.

The Flowers, the most distinguishing feature of the species, are in pendulous catkins, shaped like large acorns. Those above (next the stalk) are functionless, except for attracting insects by their scent, and consist of 10 long, mauve filaments. Those below (at the tip) are yellow and consist of a pistil surrounded by 10 stamens. The whole flower-spike is 2 inches long with a 1½ inch stalk. They first appear in February and blossom for several months onwards.

The Fruits are bunches of small, brown pods which are so twisted and contorted as to assume fantastic bundles with alternate concave and convex surfaces outwards, each section the reverse of the next. They hold about four small, flat, black seeds, pointed at one end. The pods are very persistent and remain on the tree for several months. They fall without splitting.

Uses.—The stems are made into a good quality bow, very commonly seen, and also into sticks.

Leaf

Thorn & 2 Leaf-Stalks

Pod

Seed

A Mauve Flower

A Yellow Flower.

The Flower-Spike.

DIOSPYROS MESPILIFORMIS Hochst.—*Kainya, Kaiwa.* *"Ebony."*
EBENACEAE.

This species, though typically inhabiting the banks of streams and the depths of "kurmis," is quite commonly found in dry situations, where it attains large dimensions. It is the same species as is found in abundance in the south, but while there it yields a fair proportion of black heartwood, in the north there is but a pencil or none at all. The form varies greatly according to locality.

On stream banks in the savannahs, the foliage extends, as a rule, down to ground level, or near it. In the "kurmis" great, tall, tapering stems, cylindrical or, in some cases, apparently built up of a number of stems which have joined and formed a fluted column, or, in other cases, of a number of separate stems forming a huge crown, are the types. In some cases, *e.g.*, north-east of Sokoto Province, a number of large trees occur in the middle of farm land on light sandy soil. These may reach a height of over 60 feet and a girth of over 10 feet.

The Bark distinguishes the tree from all others. It is almost black, and the scales are small, regular, even-sized and rectangular. That of very young trees is green or grey and quite smooth. The slash is salmon pink with darker flecks.

The Wood of the northern examples rarely contains any black wood. The heartwood is a mixture of shades of pink, grey and green, the predominant colour being light red. In transverse section the rings are faint waved dark lines, close together; the pores are small, open, evenly scattered, mostly single, with some small chains in the line of the extremely fine and close rays. The rays show as small red bands in radial section. The grain is close, not very straight and the wood seasons well and is hard and durable. Damage from fires to young trees often extends many feet up the stem. It works fairly well with tools but will not take nails well. The planed surface is smooth and will take a polish. The weight is 48 lbs. a cubic foot.

The Leaves are a very dark and rather dull green, the under surface with a slight sheen. The venation is very delicate and not raised on either surface, though the mid-rib is most prominent underneath. They are some 6 inches long and 2½ inches wide, with a short stalk. The young foliage is bright red, which turns brownish before the final green. The shade is dense and nothing grows under it.

The Flowers are borne in small clusters in the axils of the leaves in February. They are green and not conspicuous. Each consists of a 5-lobed calyx on a longish stalk, 5 petals set in spiral formation, one over the other and inside the closed corolla are 10 stamens attached to the petals.

The Fruit is about an inch in diameter, green at first, with the crinkly calyx much enlarged and toughened, ripening to yellow with a crisp rind, a soft, sweet, edible flesh and from 4-6 seeds. The seeds are very hard with shiny, rich red-brown coat and grey interior into which the red colour of the coat runs in streaks as seen in section. The seeds are grouped radially round the centre with thin, jelly-like partitions between each. The fruit falls entire from March onwards.

Uses.—The tree is cut into planks for various uses and the timber is very durable though a little too heavy. Canoes have been made of it. The fruits are eaten fresh by the natives.

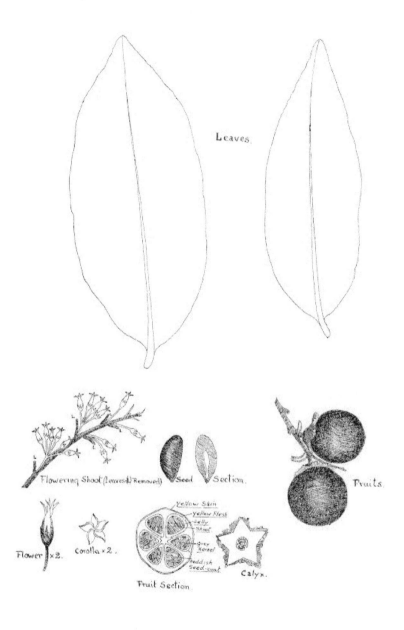

Leaves.

Flowering Shoot (Leaves Removed). Seed. Section.

Fruits.

Flower ×2. Corolla ×2.

Yellow Skin
Yellow Flesh
Jelly
Sheet
Grey
Kernel
Reddish
Seed-coat

Calyx.

Fruit Section.

EKEBERGIA SENEGALENSIS A. Juss.—*Madachin dutsi.*
MELIACEAE.

This not very common species is found in Bauchi, Zaria, S. Katsina, etc., and has not been observed above 11½° N. It inhabits the outskirts of "kurmis" or the open forests where the soil is good. Considerable quantities occur just below the 4,000 feet level on the Bauchi plateau. The similarity of leaves, flowers and fruits and its rocky habitat are responsible for its local name. It reaches 30 feet in height with a girth of 2-3 feet, rarely has any bole length over 10 feet, but as a rule a number of acutely ascending limbs forming a dense oval crown. It stands shade well and there its growth is taller and more slender and open and inclined to spread. The terminal foliage is noticeable.

The Bark is smooth and dark grey, often with marked light and dark patches, and on younger stems or branches is covered with white lenticels in vertical rows. The slash is crimson with white streaks. The bark of older trees is roughish, with close-fitting, small scales.

The Leaves are about 15 inches long, pinnate, with 6-10 pairs of opposite, or nearly, pointed, wavy-edged leaflets some 3 inches long, a dull dark green above, pale beneath, with a prominent mid-rib on the under side which is reddish, as is the stalk of the leaf. They are borne on the last foot of the twig in erect, rosette-like bunches.

The Flowers are in panicles 5-6 inches long, amongst the leaves in February and March. Each flower has 5 pale green petals, 10 stamens with black anthers and a knobbed pistil. The filaments of the stamens are united up to the anthers.

The Fruits are capsules 1-1½ inches in diameter, normally of 4 segments, but owing to the growth of one or more at the expense of the rest, the capsule is often irregularly developed. They are light brown in colour and at first sight resemble those of *Khaya* but contain brilliant red oval seeds set in a bright yellow aril, embedded in the thick walls of the capsule, in a vertical position.

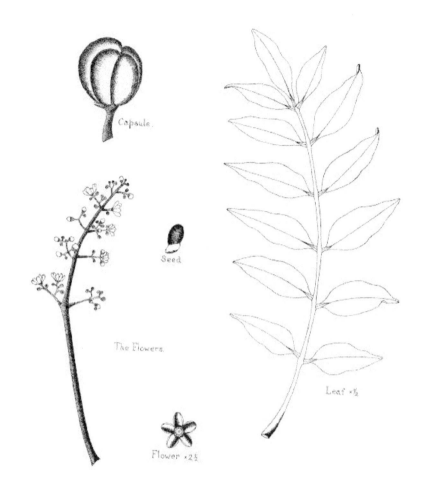

Capsule.

Seed

The Flowers.

Leaf ×½

Flower ×2½

ENTADA SUDANICA Schweinf.—*Tawatsa*. LEGUMINOSAE.

A very common species occurring everywhere in savannah forests of the more open type; usually about 20 feet high, sometimes more, with an average girth of 3 feet. It forks and branches low and has, as a rule, no length of bole and a widespreading, open crown. In full leaf it is graceful in appearance and not unlike *Parkia*. It is of no use as a forest species except from its frequency.

The Bark is light grey with very long fissures and ridges set widely apart. It peels in long strips and provides a fibre, as does that of the roots which is in local use for rope. The slash is crimson and white in thin streaks.

The Wood is a light red colour. In transverse section the rings are well marked dark lines, close together, the pores are large, open, few, widely scattered, single or in small nests or rows of two or three, the rays straight, continuous, not too evenly spaced, some just visible to the naked eye. In vertical section the pores are long, open and sparse, and the grain fairly well marked in darker red lines. The wood is not hard, easily worked, planes to a nice smooth finish and is sound and compact. The weight is 50 lbs. a cubic foot.

The Leaves, which resemble those of *Parkia filicoidea*, though they are a much lighter green, are bipinnate and about 18 inches long. The pinnae, 5 or 6 pairs, are set wide apart on the mid-rib, and there are some 15-20 pairs of inch-long leaflets, light green above and grey-green beneath, with a waxy texture. They grow densely near the twig tips and the mid-rib remains on the twig after the leaflets have fallen, persistent through the dry season.

The Flowers are in 4-5 inch spikes, whitish in colour. Each flower consists of 5 small green sepals and 10 white stamens with pistil. They appear amongst the new leaves in March, and are sweet-scented.

The Fruits are the most conspicuous feature of the species and the most ready means of identification for several months in the year, as they are very persistent, in whole or in part. Each is a long, flat, ribbon-like pod, from 6-15 inches long and over 2 inches broad with as many as 15 seeds, embossed. A rich brown in colour, the pod splits into sections between each seed, the outer cover falling away and releasing the seed which has a wing the shape of each section. The seed itself is a small, flat, brown oval, and the papery wing is veined. The seeds fall one by one, leaving the skeleton rim of the pod itself persistent for some time. The pods are a prominent and disfiguring feature of the tree, in the dry season.

Uses.—The leaves are appreciated as cattle fodder. An infusion of the bark is drunk as a tonic. The bark of stem and roots gives a fibre used as rope.

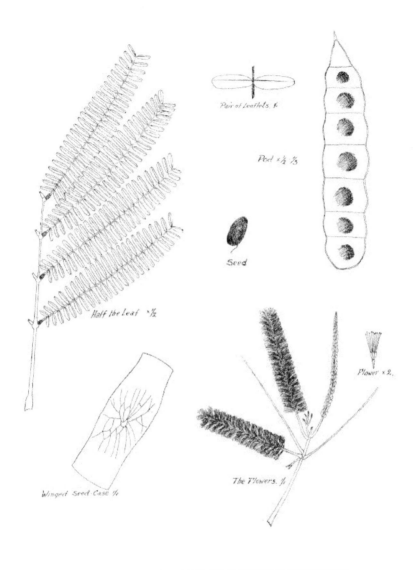

Pair of Leaflets, ½

Pod × ½ ⅔

Seed

Half the Leaf × ⅛

Winged Seed Case ⅓

Flower × 2.

The Flowers. ½

ERIODENDRON ORIENTALE Steud.—*Rimi. The White-flowered Silk-Cotton.* MALVACEAE.

This well-known tree is the tallest in the northern provinces and a height of 100 feet may be reached, with girths of 12 and 15 feet. Those who are familiar with the giant examples of the south will readily distinguish the difference in form between them and the type found in the north. In place of the clean bole up to 100 feet long with great horizontal limbs, the crown descends to within 10-20 feet of the ground and is, as a rule, very regular and shaped like a sugar-loaf, with a pointed top. The other type is seen but is not typical of the region. The branches, too, ascend at an acute angle and in the case of trees which have been lopped, a very common occurrence in the north, the new branches run up parallel with the main stem. Here and there occur branches which grow at right-angles to the main stem, so that one tree will show both types. The tree, as a rule, bears large, conical thorns, but there is a variety called "Rimi Masar" which has no thorns at all. The young tree shows a whorled arrangement of the branches. It is one of the quickest-growing trees and can be grown from poles, stuck upright in the ground.

The Bark is pale grey and smooth. In young specimens it is a bright green, with or without a mass of stout, conical thorns with round bases and sharp, black points. The slash is crimson and white in patches, the white darkening to brown.

The Wood is white with small, yellowish streaks, and is soft and fibrous when green, and brittle and inclined to be crumbly when dry. It rapidly rots on exposure to weather but will wear quite well when made up into any article, such as a stool. It is short in the fibre and has been rejected as a source of pulp for that reason.

The Leaves are truly digitate with 7-9 lobes, each 3-4 inches long and narrow and pointed. They are dark green and smooth and slightly paler beneath, with a long stalk.

The Flowers, which appear in December cover the leafless twigs in pendulous masses. They are lily-shaped, with 5 brownish petals, so densely covered with silky hairs as to appear white on the tree, especially when in bud. There are 5 stamens with rolled-up anthers and a long, white pistil. The attachment of the pistil to the ovary is peculiar. The concave base of the corolla, which sits on the ovary, is pierced with a small hole through which the pointed end of the pistil passes, but when the corolla falls entire, the pistil is prevented from remaining attached to it by a swelling of the style at the base of the corolla, on the inside. The ground may be seen littered with the fallen flowers, each containing the unattached pistil.

The Fruits are large, pendulous capsules, 6 inches long, which are black when ripe and split into 5 sections, releasing a number of small, black seeds which are embedded in a mass of silky fibre known as Kapok. The fibre grows from the inner surface of the capsule.

Uses.—The wood is used for canoes of inferior quality in places where a better timber is not handy.

The silk-cotton is used for stuffing all kinds of articles, and by the native for stuffing donkeys' saddle-pads. It is also used as a kindling for flint and steel.

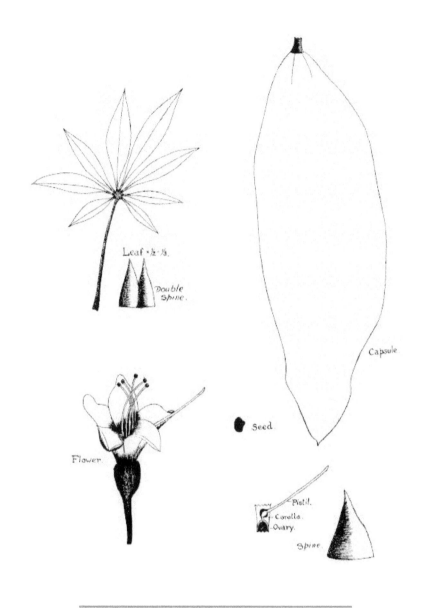

Leaf ×½–⅓.

Double Spine.

Capsule

Seed.

Flower.

Pistil.
Corolla.
Ovary.

Spine.

ERYTHRINA SENEGALENSIS DC.—*Minjiriya. Majiriya.*
LEGUMINOSAE.

This species is common locally and prefers the banks of streams, on which it grows to large dimensions. Examples 40 feet high with a girth of over 6 feet are met with and numbers may be found growing together along a considerable length of a stream. It occurs also, more or less commonly in open forest country, but does not reach the same dimensions as those near water. The larger trees have 20 feet boles and large, crooked limbs forming a wide, open crown, irregular in form.

The Bark of old trees is very rough and, in the case of those trees growing on stream banks, covered with lichen. There are long, vertical fissures and long, heavy scales. The stem of the young tree, and the branches of the old, are covered with short, sharp thorns with heavy bases, in pairs, pale brown with black points. The slash is yellow.

The Leaves are trifoliate, the lateral pair without stalks, the terminal larger with an inch stalk. They are a dull, grey-green, often blotched on the under surface. The leaflets are 2-3 inches long.

The Flowers readily identify the species and are borne in bright red, terminal spikes. Each flower is a highly modified form of the "Pea," and consists of a hood-shaped calyx and a large standard petal, folded down the middle, enclosing the long pistil and 10 stamens, some of which protrude with the pistil. They appear from November-January.

The Fruits are pods, about 6-7 inches long with embossed sections. They are sickle-shaped and dull green with a velvet covering. When ripe they split down one side and twist into fantastic bundles with the coral-red seeds attached alternately to the rims of the two sections. The vivid colour of the seeds gives the name "Coral Tree" to the species.

Uses.—The seeds are used locally by boys in playing games of chance.

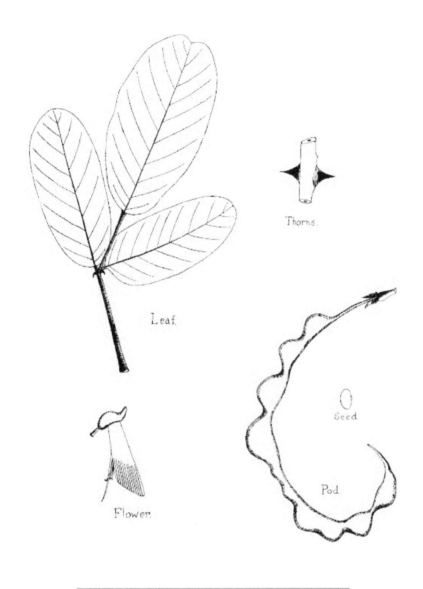

Leaf.

Thorns.

Flower.

Seed

Pod.

EUGENIA GUINEENSIS DC.—*Malmo*. MYRTACEAE.

A very familiar species which grows along stream banks as far north as 12°, while from 11° south it may be found gradually more and more able to leave these where conditions are suitable as regards soil and moisture. In Zaria and Bauchi especially it may be seen along streams to the exclusion of nearly all other species, and where it attains large dimensions up to 40 feet with girths of 6 feet. Like most species of the fringing stream belts the foliage approaches close to ground level, except where good protection is formed by the wider influence of "kurmis" when longer boles are found. The crown is dense but not high. Both the flowers and fruits are distinctive features.

The Bark is very dark, sometimes almost black or very dark brown, often lichenous with a mottled appearance. The bark is fairly smooth with long rectangular scales of little thickness. The slash is bright crimson and of fibrous texture.

The Leaves are very variable in shape and size, the true type being oval with rounded base and short pointed tip. They are 3-5 inches long and variable in breadth in proportion to a constant length. The surface is smooth with a rubbery texture, dull wax-like green. The venation is very fine, consisting of a large number of parallel laterals inclined acutely forward connected by a fine network of interlacing veins. The leaves are often distorted, the mid-rib curved or the tip of the leaf cleft.

The Flowers are in terminal racemes, regularly branched, of white flowers, each composed of a funnel-shaped 4-pointed calyx, 4 petals which are noticeable in bud, numerous white stamens which spring from the edge all round and a short curved pistil. They are scented and appear from January to March.

The Fruits are drupes in large falling clusters passing through the green, white and mauve stages before ripening to a rich purple-black. They are oval, a little over ½ inch long, with a prominent mouth. The skin is thin, the juicy flesh narrow and the thin-skinned kernel large. Birds eat them with avidity.

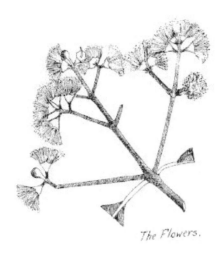

The Flowers.

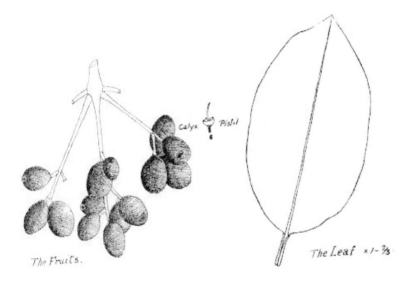

Calyx   Pistil

The Fruits.

The Leaf ×1-⅔

FICUS CAPENSIS Thunb.—*Uwar yara, Haguguwa, Farin baure.*
MORACEAE.

A large fig tree commonly found on stream banks and in gulleys, distinguished by its masses of pear-shaped figs clustered round the trunk of the tree. It grows some 30 feet high with a girth of 4-6 feet, and in form is usually tall and narrow with a cylindrical crown which extends down to near ground level.

The Bark is a light brown colour and the scales are small, rectangular and grey, in patches on the tree. The slash is light red with a flow of milky sap.

The Leaves are 3-4 inches long and some 2 inches wide, of rather unusual shape, with broad tip, unequal basal lobes and wavy edges parallel in the middle portion. They are dark green with a bluish, waxy upper surface, paler beneath, and have ½ inch, stout stalks. The foliage is dense.

The Figs are borne in dense clusters on the trunk and wood of the larger branches, much branched twigs bearing the heavy crop. They are pear-shaped, about an inch long, and smooth. They are not edible.

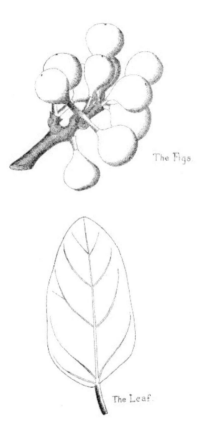

The Figs

The Leaf

FICUS GNAPHALOCARPA A. Rich.—*Baure*. MORACEAE.

A very large fig tree, common throughout the country, but not in dry localities. It demands a moist site, preferably on the banks of a stream or in a fairly well watered locality. Though giving a deep shade it is not, like the "chediya" and "durumi" planted in towns for that purpose. It occurs in gregarious clumps, self sown and the seed is distributed by birds or goats and sheep which are very partial to the figs. It grows to very large dimensions, 50 feet in height with girths of 10-20 feet. There are two common forms: the high-crowned with long bole, erect branches and rather narrow, flat-topped crown, and the heavy-crowned, short-boled tree with low, widely-spreading limbs and immense crown. This is probably, like many other trees, merely a question of age and surroundings. Flanges a foot or two high and the roots running above ground for several feet are common features. The tree puts on its new foliage very quickly and this is a brilliant dark green colour.

The Bark of young trees is a light green colour with a soft, powdery covering. That of older trees is very distinctive, grey-green, fairly smooth with grey scales here and there about the bole and light brown patches where these have fallen. The trunk presents a mottled appearance most of the year, and green, grey, light and dark brown and bluish tints are intermingled. The slash is pale pink with a flow of milky sap.

The Leaves are 4-5 inches long and 3-3½ inches broad, cordate at the base with irregular margins. They are dark green on both sides with the veins very clearly visible in lighter colour and all prominent on the underneath. Both surfaces are rough to the touch, this being the most distinguishing characteristic of the species. There is a ½-¾ inch stalk.

The Figs are borne directly behind the leaves in short clusters any time from December to March and there is a heavy crop. They are about 1½ inches long and 1¼ wide, bluntly pear-shaped with ¾-1 inch stalks. When ripe they are various shades of red and orange to almost purple, rarely uniformly coloured throughout, and covered densely with short straight hairs like plush. They are much eaten by the natives, especially by children and are certainly the best eating though much spoilt by the fertilising fly. Sheep and goats relish them.

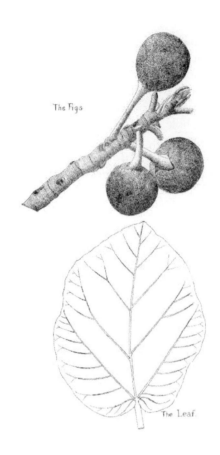

The Figs

The Leaf.

FICUS ITEOPHYLLA Miq.—*Shiriya, Shirinya.* MORACEAE.

A very well known species of fig tree distinguished by its long narrow leaves and long clusters of small figs. It grows to great sizes, over 40 feet high and with girths of 15 feet. It varies considerably in form from a tall narrow tree to one with enormous flat-topped crown 75 feet in diameter. The foliage is superficial and the shade is not very good. The bole is short, gnarled or compound, and there are many roots trailing over the ground for several feet from the base of the tree in old examples. The branches of younger trees are very erect and it is only in old age that the large spreading form is assumed. It often starts life as a parasite, the seeds being carried by birds to a crevice in another tree which it gradually surrounds and smothers. It is not unlike *F. Kawuri*, whose leaves are, however, slightly broader and whose figs are hairy and reddish.

The Bark is a creamy colour and a few grey scales are scattered here and there over it, giving it a mottled appearance. The bark is smooth.

The Leaves are 3-4 inches long and an inch wide with an inch stalk. They gradually widen from the base upwards and suddenly narrow to a point. The upper surface is dark, shiny green and the mid-rib and leaf stalk are very light green. The under surface is lighter in colour. The mid-rib is prominently raised on the underneath only. The margin is slightly waved. The leaves are arranged spirally in clusters of about 20 on the last 3-4 inches of the twigs, presenting a whorled arrangement.

The Figs are in 3-9 inch long clusters on the otherwise bare twigs, though they may be borne on the tree when it is in leaf. They are just over ¼ inch in diameter and when ripe are green with a red tint, round and the surface covered with small warts. They are eaten by birds which distribute the seeds to other trees where the tree starts life as a parasite. The crop of figs is a very heavy one.

FICUS KAWURI Hutch.—*Kawuri.* MORACEAE.

This is perhaps the largest fig tree with the exception of *Ficus polita*, "durumi," attaining a height of 50-60 feet with girths of over 20 feet. It may start life as a parasite, when it forms the usual mass of aerial stems forming a compound bole. It grows to vast dimensions, forming a sun-proof shade with its dense, regular-shaped and clearly outlined crown. It is common in and around towns. The slash is pale red-brown with a flow of milky sap.

The Leaves are large, some 6 inches long and 3 inches wide on an average, 8 inches long and 3½ inches wide being the largest. They are broadly spear-shaped with cordate base and tongued tip. The upper surface is very dark green and shiny, the venation clearly marked and prominent on the surface, the mid-rib and main lateral nerves only prominent on the duller under surface. The leaves spring from the short, blunt twigs in a rosette formation.

The Figs are small, squat, pear-shaped and with black mouths, and are borne in small clusters of 2-4 in the axils of the leaves or behind the leaves on the wood. They turn from whitish to pink with purplish tints when ripe, and birds are very partial to them.

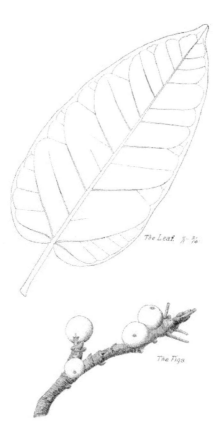

The Leaf ⅜-¾

The Figs

FICUS PLATYPHYLLA Del.—*Gamji*. MORACEAE.

A common species of fig tree, readily distinguished by the great size of its leaves and by the rusty colour of its bark. It attains a great size, over 60 feet in height, with enormous girths. The size of the large spreading limbs and breadth of crown are a feature. The lower limbs are horizontal and of great length and the crown is either flat-topped or round. There are small root flanges and often great lengths of root above ground. Though the leaves are few in number their size makes the tree give good shade. Life is started epiphytically.

The Bark, especially of younger trees and the branches of old trees, is rust red and there are large, scattered or patchy scales on the bole, which are light grey. The slash is dull pink, with a flow of milky sap.

The Leaves are very large and dark green. Twelve inches long and 8 inches wide, the upper surface is velvety with hairs and the veins show a pink tint. The under surface is lighter and the veins so prominently raised as to resemble rubber piping. The base is cordate and the tip pointed. The new leaves spring from the tips of the twigs above the figs.

The Figs are borne in clusters a foot or more in length and the crop is often a very heavy one. They ripen in December or January, or in places later, and the new foliage grows above them. The twigs on which the figs are borne are an inch thick with blunted tips. The figs themselves are an inch in diameter and have a 1½ inch stalk. They are reddish in colour with a soft skin and are covered with red warts.

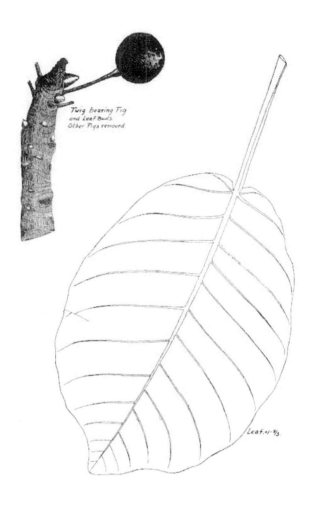

Twig bearing Fig
and Leaf Buds.
Other Figs removed.

Leaf. x¹⁄₃.

FICUS POLITA Vahl.—*Durumi.* MORACEAE.

This is one of the commonest of the fig trees, chiefly occurring in towns where it is planted for its shade. It is the densest of all the shade trees and readily identified by its very large, spherical or flat-topped crown of darkest green, and heart-shaped leaves. It branches not above 10 feet from the ground and innumerable straight branches extend in every direction to form the dense, superficial crown. In form it differs from *F. Thonningii* in being flatter topped and not so tall, the crown being umbrella-shaped. The trunk of an old tree may be of great size and is built up of many aerial stems formed from the growth downwards of the aerial roots which anchor themselves to the soil or hang in festoons from the branches. A network of roots may often be seen spreading several feet round the base of the tree.

The Bark is smooth and light in colour and creeping down it or suspended from it are the aerial roots, red at the tips. The bark yields a copious flow of milky sap.

The Leaves are heart-shaped with a prominent tip and a waved edge, are dark green and shining above, smooth beneath, with the main veins raised on both surfaces. They have a 2 inch stalk and are 4-5 inches long and 3-4 inches wide. They are pendulous.

The Figs are single or in pairs on the wood of the branches and are very numerous, clustering all round the branches. They are about 1¼ inches in diameter and have square, twisted stalks over an inch long. They ripen in January.

Uses.—It is planted, from poles, solely for its shade and in old towns trees may be found with a shade diameter of 100 feet.

The figs are sometimes chewed by the native but not swallowed.

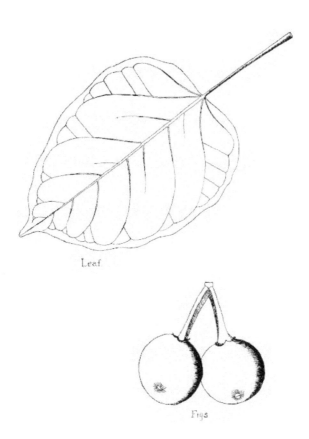

Leaf.

Figs

FICUS THONNINGII Blume.—*Chediya*. MORACEAE.

This species, somewhat similar to *F. polita*, is also commonest met with in towns, where it is grown for its shade. In form it differs from *F. polita* in that it reaches a greater height, 60 feet being not unusual, and has, as a rule, a spherical, rounded-topped crown. Numerous straight branches ascend from a large, short stem and, bearing their leaves at the ends, form a dense, superficial crown. The stem itself is, as in the case of *F. polita*, composed of a large number of aerial roots which have grown downwards from the branches and entered the ground, eventually combining with the stem to form a gigantic whole. Where the main stem branches, a dense mass of small aerial roots is often in evidence. The species is commonly seen in the act of strangling another tree in some fork of which the seed has originally lodged and about which it has twined its aerials till the host is completely enveloped. It is a fast growing species and readily propagated from a branch length placed upright in the ground.

The Bark is light grey and smooth and a copious flow of milky sap will pour from a cut.

The Leaves are dark green and slightly shining on the upper surface, on which the venation is prominent. They are about 5 inches long and 2½ inches broad, with 2 inch stalks.

The Figs are very numerous and are under ½ inch in diameter, in dense clusters amongst the leaves. They are reddish in colour with a warty skin. They ripen about February, and are not edible.

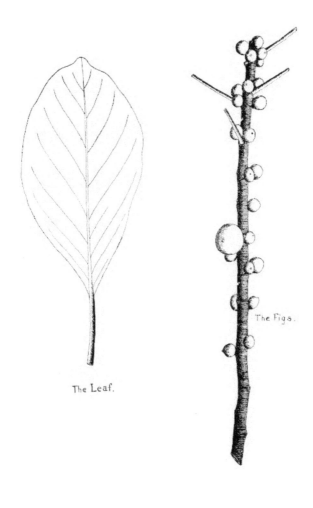

The Leaf.

The Figs.

FICUS VALLIS-CHOUDAE Del.—*Dulu*. MORACEAE.

A large fig tree, most commonly found along the banks of streams and only away from them farther south in moister conditions. It reaches a height of 50-60 feet with very large girths. The bole is short, not above 10 feet and the limbs, widespread, form a large open crown with superficial foliage of very large leaves. Both the leaves and the large, solitary figs are distinguishing features. It is very common indeed in Bauchi province.

The Bark is light grey with large polygonal scales that leave light patches. The slash is pink, with a flow of milky sap.

The Leaves are large, averaging 6 inches long by 5 inches broad, but exceeding these dimensions by some 2 inches in some cases, with a stalk some 2 inches long. The base is slightly cordate and the margin broadly waved or bluntly toothed at wide intervals. The venation is prominent underneath and flat above. The upper surface is dull, dark, smooth, but not shining, the lower side slightly paler.

The Figs are solitary and large, almost up to 2 inches in diameter, flattened or pear-shaped, with a slight, softly hairy coat, and traces of longitudinal lines. The mouth is large and protrudes very slightly. They may be found from February to June.

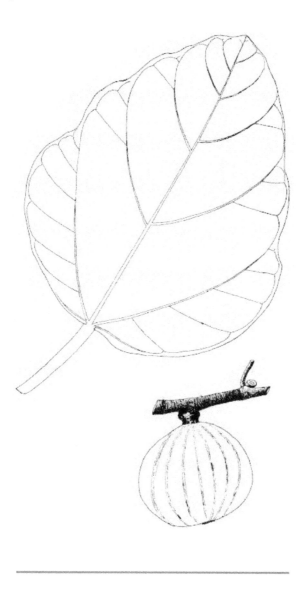

GARDENIA ERUBESCENS Stapf & Hutch.—*Gaude*. RUBIACEAE.

A common and conspicuous shrub of the savannahs whose only claim to importance lies in its frequent occurrence in large numbers over wide areas of the country. It rarely exceeds 10 or 12 feet in height as against the 20 feet to which its congener, *G. ternifolia* may attain. It is often gregarious, occurs in rich or poor and rocky soil and is more or less a component of all types of savannah. It branches from ground level and forms a round bush. The branches are twisted and bent, stout, blunt and springy.

A comparison with *G. ternifolia* shows *G. erubescens* to have larger, paler and brighter green leaves, with less crinkled margins; crisp, not fibrous fruits of more uniform and oblong shape; stouter, blunter and browner twigs, and a more shrub-like growth.

The Bark is smooth, grey-brown, with a powdery surface and small, thin scales which leave lighter scars. The slash is yellow, with green edges.

The Wood is yellow, hard, crisp under the axe, planes well, seasons well, is sound and clean and weighs 50 lbs. a cubic foot. In transverse section the rings are clearly visible as fine white lines, the pores are minute, open, evenly distributed and more numerous in the light annual rings. The rays are extremely fine, rather far apart and wavy.

The Leaves are about 5 inches long and 2½ inches broad, bright green and shiny on the upper surface, lighter beneath, with the mid-rib raised on both sides. The margin is wavy or not, and some leaves are narrow at the tip and others blunt and flat. They do not form as marked rosettes as those of *G. ternifolia*.

The Flowers are large, white, tubular, 6-petalled blooms, 3 inches in diameter with a very long, narrow corolla tube from which protrudes a ribbed club-like stigma, surrounded by 6 stamens consisting only of long anthers attached by their centres to the corolla. They are very highly scented and are conspicuous from their size, colour, number and perfume about December.

The Fruits are fleshy, potato-like, yellow, with a smooth surface and variable in size and shape, though usually long. The flower-parts are persistent at the top. The firm, juicy flesh surrounds a number of small, flat, yellow seeds in a bright yellow, pithy pulp, and is sweet and edible. They are eaten by antelopes.

Uses.—The fruits are eaten fresh or made into a sauce as an ingredient of soup.

The whole plant is cut for fencing farms as protection from goats, etc., and the twigs are stuck on the tops of compound walls.

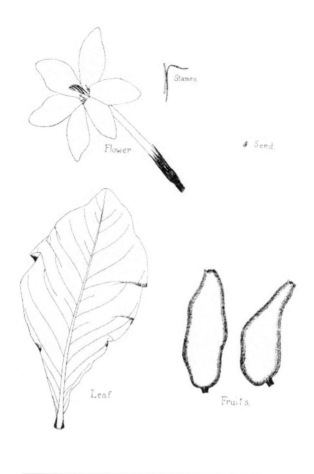

Flower

Stamen

Seed

Leaf

Fruits.

GARDENIA TERNIFOLIA Thunberg.—*Gauden kura.* RUBIACEAE.

This species, similar to *G. erubescens*, is found in the same situations, but is less plentiful as a rule. It may be seen mixed with the other species or over considerable areas by itself. It is, or can be, a taller species, occurring as a tree about 20 feet high with a less spreading habit, and a flat-topped crown. It branches from just above the ground and the branches are crooked and very springy, but not so stout and blunt as those of *G. erubescens*, in fact, the straight twigs of this species are very distinctive.

A comparison with *G. erubescens* shows *G. ternifolia* to be a larger or taller tree, less shrub-like; to have greener bark; to have smaller, darker, rougher and more crinkled leaves with pointed tips, and to have fibrous, grey-green fruits which vary exceedingly in size and in shape.

The Bark is smooth and grey or green with a powdery surface, and there are small, thin, grey scales which leave greenish-yellow scars. The end of the twig is covered with knobs of soft green cork. The slash is yellow, with green edges.

The Wood is yellow, close-grained, tough and hard to cut with the axe.

The Leaves are about 4 inches long and 1½ inches wide, dark green and slightly rough on the upper surface with a white mid-rib raised on both sides. The margin is crinkly and the tip pointed. They spring in little rosettes from the ends of the twigs.

The Flowers vary in size; are generally smaller than those of the other species but may be larger, 3½ inches across and with a corolla tube the same length. They are white and the under surface of the petals is very shiny. They have 6 petals, 6 stamens, consisting only of long anthers attached to the mouth of the corolla round a clubbed and green-ridged stigma. The calyx is irregular and a very bright, shiny dark green. The flowers are highly scented and turn yellow when fertilised. They appear about December and are very conspicuous by their size, colour and perfume.

The Fruits are very variable in size and shape, but are always of the same consistency, grey-green on the outside with a fibrous flesh that surrounds a number of small, round, flat, yellow seeds in a firm pith. They may be round or oval, ribbed or plain with roughened surface, covered with lenticels, and from 1 to 4 inches long. They may be found on the tree during the greater part of the year, and fall entire.

Uses.—The fruits are not edible and are used in the making of black cosmetic (katambiri) for decorating the faces of women.

The tree and its branches are used for fences (shingi) round farms and also for sticking on the tops of compound walls.

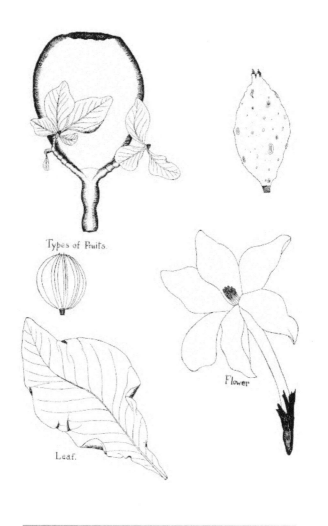

Types of Fruits.

Leaf.

Flower

GREWIA MOLLIS Juss.—*Kurukubi.* TILIACEAE.

A large shrub or small tree growing in higher savannah forest and not occurring in dry latitudes. It is especially partial to granite soils and hills. It has no very definite form, being either a dense shrub 12-15 feet high or a small, shapeless tree up to 20 feet, with crooked branches and a mass of fine twigs. It likes the shade of high forest and is found under the shelter of large trees. The star-shaped yellow flowers and irregularly toothed leaves are the conspicuous means of identification.

The Bark is black and rough, deeply fissured and fibrously scaled. It presents a rough, shaggy appearance.

The Leaves are alternate on the long shoots and assume one plane with their surfaces to the light. Especially in the shrubby forms is this most marked. The leaf is 3-5 inches long and 1-1½ inches broad, borne on a short stalk and with the edges acutely and irregularly cut into teeth of all sizes. There are three main nerves, the mid-rib boldly branched and the laterals less so. All the nerves point well forward, and are prominent beneath. The basal lobes are unequal. The leaf-stalk is dorsally flattened. The upper surface is a dull, blue green, the under surface paler. The leaf-bearing shoots are oval in section.

The Flowers appear in April and persist till well on in June, the fruits almost ripening before the last flowers are over. They are in axillary clusters, 2-3 on a stalk and 2-4 stalks in an axis. They are ½ inch in diameter, bright yellow, with 5 long, narrow yellowish sepals, 5 small, round yellow petals and a mass of yellow stamens surrounding a knobbed pistil.

The Fruits are blackish when ripe, green with a pubescence before ripe, 2-lobed, with a cell in each lobe. They are edible.

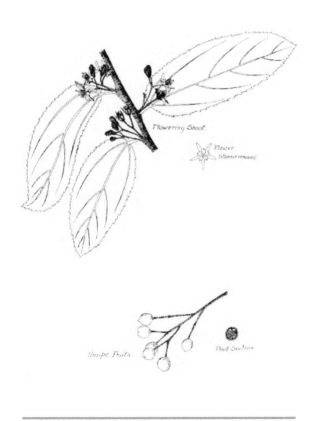

Flowering Shoot.

Flower. Stamens removed.

Unripe Fruits.

Fruit Section.

GUIERA SENEGALENSIS Lam.—*Sabara*. COMBRETACEAE.

A shrub, or, very occasionally, a small tree, which covers wide areas of otherwise barren land in the Bush savannah. In north-west Sokoto, for example, there are miles of this scrub covering what were large areas of cultivation at one time. It also covers the bare rocky slopes and summits of the flat-topped hills. On these it is associated with the smaller *Combretum* species "geza." In appearance it is a bushy plant with several stems, dusty grey-green in colour from a few feet to about 15 feet in height. It is most conspicuous in fruit. The stems are very subject to a gall caused by a grub whose brown excretions fill the cavity in the gall. A small black and brown ant is found covering the plant at times.

The Bark is grey, and that of the young stems and branches is brown and covered with loose, brown fibres running vertically.

The Wood is cream-coloured with a reddish tinge. In transverse section the rings are darker bands, the pores are very small, single, in nests and in festoons, the contents closing most of them; the rays are very fine and evenly spaced, close together. The grain is coarse and twisted and the wood is tough and of no practical value. Weight 55 lbs. a cubic foot.

The Leaves vary much in size, being from 1-2 inches long and ¾-1½ inches wide. They have a slightly cordate base, a pointed tip and are broadly oval. The stalk is ¼ inch long. They are in pairs and opposite. They are a dusty grey-green, but a fresh green when new, and are rather dry and leathery in composition.

The Flowers, which are found at most times of the year, are in greeny-yellow, spherical heads about ½ inch across, enclosed at first in a sheath of bracts which splits open first into two parts and finally into four, which bend right back and persist in the fruiting stage. The head of flowers is on an inch long stalk in the axil of a leaf and bears a pair of tiny bracts about half-way up. Each flower has a 5-lobed calyx, 5 slender petals and 10 long stamens round a straight pistil. The whole is covered with minute black dots which are raised.

The Fruits are 1½ inches long, slightly curved, pointed at both ends, extremely hard, and covered with long silky straight hairs. In section they are 5-lobed, occasionally more, and contain one long seed. They may be seen for several months in the year, chiefly in the dry season, and are very conspicuous with their silky hairs. They have often a pronounced pink tint.

Uses.—The shrub is burnt round cattle, sheep and goat camps to keep off the flies and as a remedy for colds in such herds.

The leaves, concocted with water, are a medicine for internal complaints, a preventive of leprosy and, applied externally, a cure for skin irritations. They are also drunk by women after child-birth. The leaves are often added to food to prevent indigestion.

It is commonly cut for fencing farms against herds.

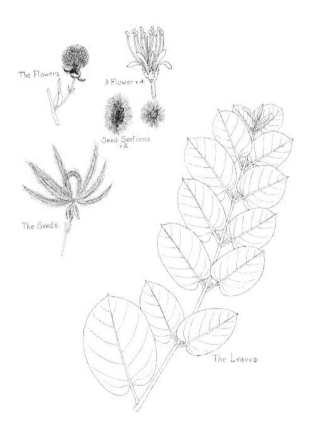

The Flowers

A Flower × 4

Seed Sections × 2

The Seeds

The Leaves

GYMNOSPORIA SENEGALENSIS Loes.—*Kunkushewa, Namijin tsada, Mangaladi, Bakororo.* CELASTRACEAE.

A shrub or small tree, occasionally as tall as 25 feet with a girth of 2 feet, but more generally a shrub found in large numbers in open secondary growth after cultivation. It does not grow in the far north and requires well-watered and more loamy soils than are found above 13° N. Sometimes it exceeds 50 per cent. of the vegetation on old farm lands, and such land being burnt every year the tree increases very slowly in height and broadens into a compact shrub. The crenate-edged leaves with bright red stalks, the small sharp thorns, small white flowers and globular fruits are distinctive characters.

The Bark of old trees is dark or light grey and covered with very small, close-fitting rectangular scales. That of the shrubs is pale-grey, sometimes almost white, and smooth. The slash is crimson.

The Thorns are in the axils of the leaves and in their axils is a bud. On the new shoots they are green with crimson bases, ¼-½ inch long, very sharp and straight. On old wood and twigs they are brown and very strong.

The Wood is whitish, hard, straight grained, sound and clean and weighs 45 lbs. a cubic foot. The rings are visible as extremely fine lines of alternate hard and soft tissue, the pores and rays both invisible except under good magnification. Planes and saws well but splits in seasoning.

The Leaves vary a great deal in size and shape, but the type is obovate or spoon-shaped with a tapering base and a rounded, broad tip, with or without a cleft of variable depth. Often the tip of the leaf is very deeply cleft, the edge is finely crenate or serrate, the surface smooth, the laterals hardly visible except when held up to the light. The mid-rib is prominent beneath, the laterals waved, the stalk crimson, the length up to 4 inches and breadth 2½ inches. The leaf droops in the sun and becomes limp in texture.

The Flowers, from December to February, are in small cymes in the axils of the leaves, on the old wood or new shoots. On the new shoots there is generally one cyme in each axil, on the older twigs there may be four separate cymes. Each flower is ⅙ inch in diameter, with a very small 5-pointed calyx, 5 white, oval petals, 5 minute stamens and an ovary with a bifid, branched style. There are male and female flowers on the same tree, the female having no stamens.

The Fruit is a capsule, ½ inch in diameter, with 3 cells each containing 2 seeds. They are conspicuous in clusters, turning from pale green to red.

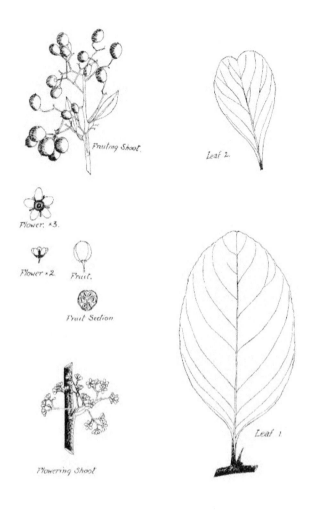

Fruiting Shoot.

Leaf 2.

Flower. ×3.

Flower ×2. Fruit.

Fruit Section

Flowering Shoot

Leaf 1

HANNOA UNDULATA Planch.—*Takandar giwa.* SIMARUBACEAE.

A medium-sized tree which abounds, though not further north than 12° and very local, in Tree savannah and the better type of Bush savannah. Its bark and leaves, once recognised are a ready means of identification. It attains a height of 35 feet with girths of 3-5 feet. It grows erect, with vertically ascending branches which form a high, narrow crown. A bole-length up to 10 feet, but not often more, is common. It is very subject to attacks by a caterpillar which skeletonises the leaves and covers the twigs with a lot of silk.

The Bark is dull, grey with long vertical ridges and fissures of very soft, thick cork. A curious effect, due to the absence of marked scales, is that the bole appears to have no bark at all, but resembles a dead tree whose trunk has turned grey with exposure. The slash is reddish.

The Wood is pale yellow. In transverse section the rings are faintly visible, the pores are mostly in little chains of 2-6 or so, having the appearance of being one long pore divided by partitions, the groups running in radial direction between the rays. Thin lines of soft tissue connect the groups. The wood is soft, saws and planes very easily, with little picking up. The wood requires well seasoning as it is liable to mould and rot. The weight is 35 lbs. a cubic foot.

The Leaves are pinnate, up to 12 inches long with 3 or 4 opposite pairs and a terminal leaflet. The leaflets are far apart and on slender stalks an inch or more in length. They vary in size on the one leaf, up to 2 inches long, excluding the stalk, and almost as broad as long in the case of the lowest pair and the terminal leaflet, with broadly cleft tips and abruptly tapering bases. The margins are sinuous, the mid-rib prominent on both surfaces and the lateral grooved on both surfaces. They are a dull, dark green above and a yellowish green beneath. The leaflets are almost at right angles to the main stalk.

The Flowers appear from October to November and are in large, loose panicles, some 12 inches long. Each is about ⅜ inch in diameter with a small cup-shaped calyx, 6 yellowish petals, 12 stamens, the lower half of the filaments being covered with hairs, and a short blunt pistil on a 6-part ovary. The flowers are scented.

The Fruits are formed by the growth of the several carpels, one or two of which generally crowd out the rest and grow at their expense. Rarely more than two reach maturity and occupy their respective positions on the stalk with their angular side facing the centre. Each is ¾ inch long and half as wide, a flattened or angled oval, with a brown or blackish wrinkled or lined skin, a hard shell and white kernel with light brown coat. Only a small proportion, as a rule, reach maturity, and these are very persistent on the tree, falling with the first rains 3 or 4 months later.

Uses.—The wood is used for all purposes that *Bombax buonopozense* is, namely basins, stools and cattle troughs, and occasionally drums.

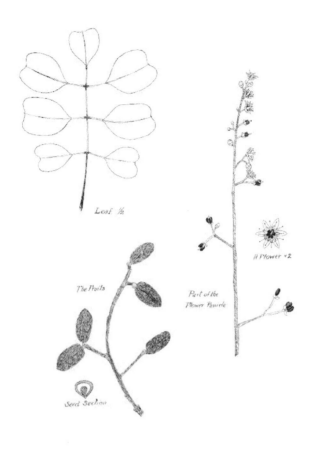

Leaf ½

The Fruits

Seed Section

Part of the
Flower Panicle

A Flower ×2

HYMENOCARDIA ACIDA Tul.—*Jan Yaro, Jan iche.*
EUPHORBIACEAE.

A small, erect tree up to 25 feet in height and 2 feet in girth. It is very widely distributed from quite near the coast up to 12° N. As most commonly seen, in open forest, it is not above 15 feet, erect, with ascending, crooked branches and a close, compact little crown. In this form it occurs in great quantities in secondary growth and is very familiar with its pale reddish stem. Older trees may have an 8 feet bole with a compact rounded crown. The male and female are on different trees (dioecious), both very readily distinguished, the first by its red catkins, the second by its red, heart-shaped, double-seeded, winged fruit. The male is often called "taramniya" by the native, owing to the resemblance of its flower-spikes to those of *Combretum* species.

The Bark is a light reddish or orange colour, sometimes almost white, with a dusty covering up to the extremities of the twigs. Here and there on older trees are a few scattered grey scales. The slash is a dull pink with a thin green edge.

The Wood is pinkish, hard, 55 lbs. per cubic foot, splits in seasoning, planes well with the grain, coarsely against it, exhibiting a bright sheen in cross section, the rings well marked fine dark lines, the pores very small and evenly scattered along the extremely fine rays in chains. The wood darkens considerably on exposure, being almost orange in transverse section.

The Leaves are some 3-4 inches long and 1-1½ broad, rounded or narrow at the base, bluntly pointed at the tip, which is variable, with ½ inch stalks. The mid-rib is prominent beneath. The leaf has a tendency to fold up along the mid-rib, especially towards the tip which is often recurved. The colour is at first a most delicate green, conspicuous from a distance but the leaf turns dark, toughens and loses the few hairs it had when young. They are arranged spirally round the twigs and stand erect.

The Flowers, male on one tree and female on another, appear in February and March. The male are in spikes, which are at first, in bud, short and bright red, gradually lengthening and turning whitish as the stamens open, finally becoming pendulous and as much as 4 inches long. Often they are in great masses and flies and small insects haunt them. Each consists of a small 5-lobed red calyx, no petals and 5 radiating stamens joined below into a short column, the anthers large, whitish and with a bright yellow gland at the tip beneath. There is a rudimentary red ovary with a bifid tip. The female flowers are almost invisible at first but grow into sight very rapidly after fertilisation. They consist of a calyx with long red sepals, varying in number from 5-8, and an ovary with two styles which become very prominent and grow long. These are wrinkled and bent or twisted and beneath them the ovary enlarges till it is about an inch long and 1¼ inches broad, gradually changing from the limp green to the dark red double-seeded, winged fruit so familiar in masses and often persistent till the new leaves of the following year.

The Fruits, as described above, ripen and split into their two cocci, each containing one large flat seed. The seed portion of the coccus is net-veined and the wing portion longitudinally veined.

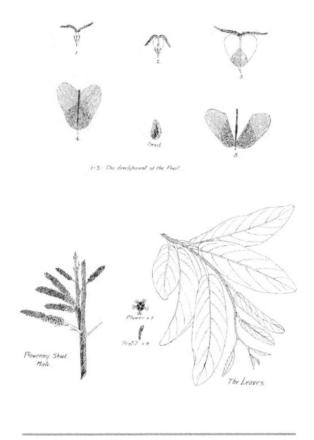

1-5: The development of the Fruit

Flowering Shoot. Male.

Flower x 6

Pistil x 4

The Leaves.

HYPHAENE THEBAICA Mart.—*Goriba. "Dum Palm."* PALMACEAE.

This very common palm is found all over N. Nigeria and is distinguished from all others by its branching habit. It thrives in dry country, farther north than any other palm. The height is some 30 feet and the girth about 3 feet. The main stem branches into two and each of these may divide again and even a third time. In agricultural land the ground is densely covered with the seedlings which are cut down each year by the farmer.

The Bark is smooth, the rings of the leaf sheaths clearly marked except at the base of old trees where the stem has the appearance of decay and an uneven surface.

The Wood is very fibrous and has no local use beyond that of roof or door supports. It has been used as fuel.

The Leaves are about 4 feet long, the segments inverted V-shaped, about 18 inches long, with smooth margins, united for a few inches of their length only. The stalk is curved back at the tip and the segments at the base are higher up on one side than on the other. The stalk is heavily armed with black spines and the sheath is divided at the base, remaining clasped to the stem for some time.

The Flowers which appear in March, when the fruit is ripe or fallen, are male and female on different trees. The spadix of both sexes is similar, up to 4 feet long with 2-3 spikes rising from the small branches at intervals along the spadix. The males are small green flowers with 3 sepals, 3 petals and 6 stamens, massed on the spikes, often spirally in rows up the spike. The females have very short stalks and have 3 sepals, 3 shorter petals, 6 rudimentary stamens and a 3-lobed ovary.

The Fruits are 2½—3 inches long and 2-2½ inches broad, light brown when ripe, reddish-brown before ripe, roughly showing 3 lobes on the exterior, smooth and hard. They contain a large, hollow kernel, very hard. The fruit ripens in March.

Uses.—The leaves are used for mats, hats, baskets, fans and various plaited articles. The unripe kernel is eaten raw and the rind of the fruit is used for making sweets. Elsewhere the kernels have been used for buttons.

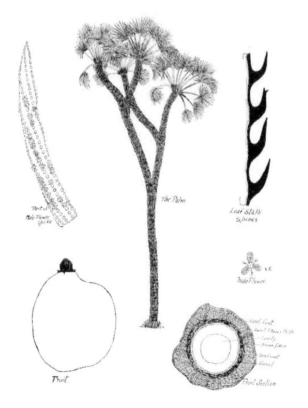

ISOBERLINIA DALZIELII Craib & Stapf.—*Fara doka.*
LEGUMINOSAE.

This species is very similar to *I. doka*, the outstanding difference being its generally paler colour in all its parts, especially in the leaves which are covered with a grey bloom. Its habit is the same and its form similar, though it differs in having a higher and rounder crown when small, the leaves not drooping so markedly as those of the other species. The bole is longer and the tree attains a greater height than does *I. doka*. A height of over 60 feet with a girth of 8 feet is not uncommon in forests of this species as well as in single examples. Long, clean boles with short, rounded root-flanges and high, flat-topped crowns are typical. It regenerates with the same profusion and is as susceptible to fire as *I. doka*, hollow stems in old trees being the general rule. This species seems to prefer stony and shallow soils and over large areas of such it grows pure, often to the total exclusion of *I. doka*.

The Bark is light grey, smoother than that of *I. doka*, and has rather large, oval scales which leave light brown scars. The slash is pale crimson.

The Wood is light red in colour with open grain of this colour and silvery streaks. The sap is dirty white. Weight 50 lbs. a cubic foot.

The Leaves are large, very variable in size and averaging some 18 inches. They have usually 7 leaflets covered with a soft grey bloom which gives them a bluish appearance. They are rounder, paler and softer in texture than those of *I. doka*, and the bloom is particularly marked in young leaves which pass through the same grades of colouring from pink to green as do those of *I. doka*.

The Flowers are in large panicles and appear in December. Each is composed of 5 irregular white petals and 5 narrow, pointed sepals, and 10 long stamens, the whole enclosed in a blackish-green, hard, rounded case which splits into two halves in the axil of which the flower sits.

The Fruit is a long, broad, flat pod, covered with a green velvet. It is about 12 inches long and 3 inches wide and splits with a report, the two halves curling up and the seeds being violently projected. This occurs in May and June.

Uses.—The young poles are used for local building purposes, but they are very soft and liable to attacks from white ants and borer-beetles and last only one season. Larger trees are locally cut into planks, and provide an inferior timber.

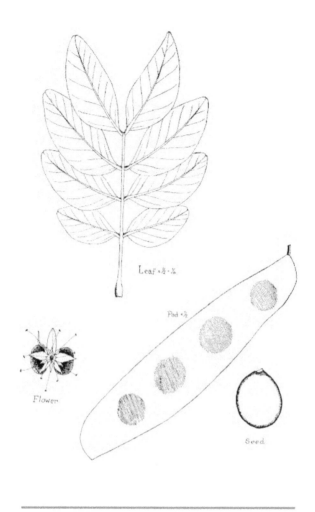

Leaf × ½ - ¾

Pod × ½

Flower

Seed

ISOBERLINIA DOKA Craib and Stapf.—*Doka*. LEGUMINOSAE.

This species, with its congener *I. Dalzielii*, is a type tree of the Tree savannah and the most common of all trees in that formation. It occurs as high-crowned forest over many hundreds of square miles of the northern provinces, though it does not extend to the extreme north. It is found in all stages according to age, from the dense covering of the cleared farm land to the 40-50 feet high-crowned forests with little undergrowth except that of its own shoots or seedlings. It grows long straight boles, usually forking at 10-15 feet, and its wide-spreading limbs form a high open crown which closes the roof of the forest. Trees in a 25 feet forest have girths of, as a rule, not more than 2-3 feet, and the crown is ¾ of the total height, and narrow, the branches ascending vertically. A girth of 5-6, or more, feet is common in trees 50 feet high. The species is very susceptible to fire and the majority of large trees have hollow centres from this cause. The stem is often a mere shell, filled with the earth of white ants or with a copious volume of reddish sap which flows out and bubbles up from the base when felled. This species seems to be more exacting as to soil conditions than *I. Dalzielii*, the latter predominating on higher or more stony ground, the change from one species to the other often being very marked. As a rule the species are mixed indiscriminately, especially on level country.

The Bark is dark grey with large, even-sized, rather shaggy scales. It is darker and rougher than that of *I. Dalzielii*. The slash is light red, the sap sticky.

The Wood is red, varying from bright red to greyish-red, with long, silvery streaks. In transverse section the pores are large, in rows, festoons and small groups, and the rays are extremely fine and close together, showing as small red bars in radial section. In the plank the pores are long, open red lines, the grain often being strikingly waved. The sapwood is silvery with a reddish tinge. The timber saws and planes well, with a smooth finish and no polish. The weight is 50-55 lbs. a cubic foot.

The Leaves are large and pinnate with an average of 7 leaflets, dark-green and shiny. The young leaves are bright red and glistening and pass through every shade from this to the ultimate brilliant green of the new mature leaf. This is most noticeable after a fire when the forest is a blaze of colour. The leaves vary very considerably in size and average some 15-18 inches, those of smaller trees being much larger.

The Flowers, which are similar in both species, are in large panicles, and each consists of 5 white, uneven-sized petals and 10 long stamens. The sepals are small, narrow and white, and the whole flower is enclosed in a hard, round, blackish-green sheath which splits into two portions in which the flower rests. They appear in February.

The Fruit is a large, broad, flat pod some 12 inches long and 3 inches wide, a dull green in colour with a smooth surface. It contains some 4 or 5 large oval, flat, pale-brown seeds which are distributed by the violent splitting of the pod, the two halves separating with a report and curling up into a spiral.

Uses.—The young trees are used for house building poles, but only last for a season as they are very soft and readily attacked by white ants and borer-beetles. Large trees are cut into planks for local use, but the timber is very inferior.

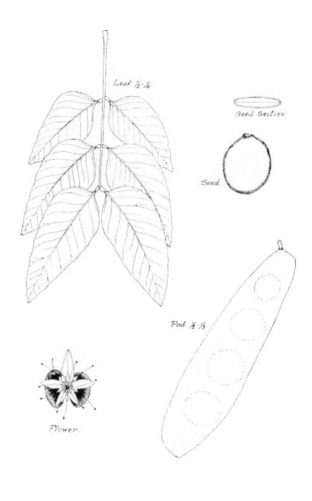

Leaf ½·¼

Seed Section

Seed

Pod ½·⅛

Flower.

KHAYA SENEGALENSIS Juss.—*Madachi.* *"Mahogany."* MELIACEAE.

This common and well-known Dry Zone Mahogany occurs more or less plentifully all over the north and has a distribution from the rain forest to the French boundary. Its habitat is the banks of streams but its soil and water requirements are comparatively modest and any little valley or moist hollow is sufficient. It abounds in the more northerly "kurmis," and on the edges of "fadammas" often shows a gregarious habit, clumps of a dozen or so growing close together. Normally it is a small tree some 50-60 feet high with a girth of 6-8 feet, but 70-80 feet and a 10 feet girth are common. Usually not more than 20 feet of its height is clean bole, but no rule can be laid down, as the stem may or may not divide into 2 or 3 large limbs which ascend vertically and form the characteristic open, wide-spreading crown. In the gregarious clumps the tree may have as many as 5 or 6 of these false stems and form an enormous crown. The bole and main limbs show a wavy habit, especially noticeable in smaller trees. The base of the bole is often much swollen by the repeated bark chipping for the bitter tonic.

The Bark is dark grey and covered with small, thin scales. A red sap exudes from the bright crimson slash.

The Wood is a deep red-brown, generally with a pronounced purple tinge which distinguishes it from *K. grandifolia*, the typical mahogany hue being absent. In transverse section the rings are indistinct colour variations; the pores are small, open, single and scattered about between the long wavy rays which are close together and visible to the naked eye. In vertical section the pores are open and dark coloured, the grain showing as waved bands of varying depth of colour speckled with the pores. The wood is very hard, sometimes badly cross-grained and tough, but at others sound and fairly straight, enabling it to be sawn and planed fairly easily. It is very apt to pick up, but the finished surface takes a good polish. The pore contents glisten. The sapwood is grey with a purplish tinge. The weight is up to 60 lbs. a cubic foot.

The Leaves are pinnate with 4-6 pairs of leaflets. The leaves of young trees will bear as many as 10 leaflets. The leaf is bright and shiny when young but darkens and dulls and the greyish under surface is typical.

The Flowers are in lax panicles amongst the end leaves and are found from December to April, earlier in some parts. The panicles are 6-8 inches long and the small flowers are white with 4 sepals and petals and 8 stamens whose filaments are united to form a crown round the knobbed pistil. The tree often flowers out of season.

The Fruits are capsules, grey, and erect and conspicuous on the top of the tree. They are from 2-2½ inches in diameter and split into 4 sections in each of which the flat, winged, mahogany-brown seeds are tightly packed one above the other. They ripen from March onwards of the next year, the seeds falling from the split capsule on the tree.

Uses.—The wood for furniture, canoes and mortars.

The bark as a bitter tonic after boiling with water, and powdered as a cure of sore backs of horses.

The leaves are gathered for camel and cattle fodder.

The seeds, dried, fried, beaten up and boiled to extract the oil, are used for anointing the body, by pagans.

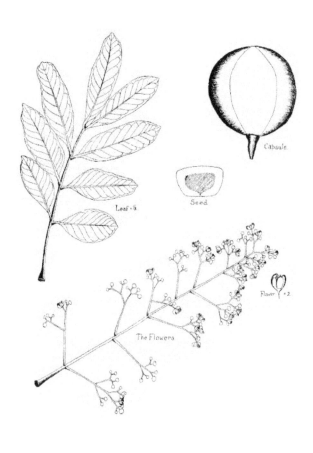

Leaf ·½.

Seed

Capsule.

Flower ✕ 2

The Flowers

KIGELIA AETHIOPICA Decne. *var.* **BORNUENSIS** Sprague.—
Rawuya, Rahaina, Nonon Giwa. "Sausage Tree." BIGNONIACEAE.

A large tree up to 50 feet in height with a girth of 6-7 feet. It occurs in the north mostly in "kurmis" up to 11° N. or more. It is not a savannah tree but has penetrated a great distance into them by way of the streams. It is low-branching or forking, and has large limbs forming a high or wide crown of great spread and dense foliage. Either in flower or fruit it is unmistakable.

The Bark is grey or light brown with small, soft, corky scales, rather scattered and leaving lighter brown patches.

The Leaves are 12-15 inches long, pinnate, with 3-5 pairs of opposite leaflets and a terminal leaflet. The leaflets are obovate with broad tip and small point, and are increasingly longer and larger from below upwards, the lowest pair about 2½ inches long and 1½ inches broad, the upper pair 4-5 inches long and 2-2½ inches broad, the terminal as broad, but not so long. The nerves are prominent beneath but grooved above. The colour is light green and shining, smooth beneath. The margins may be serrate, the teeth quite sharp, or may be quite entire.

The Flowers are in panicles up to 4 feet in length hanging from the branches in February to April. The short flower stalks are curved, the calyx is fleshy, 4-lobed and pale green, and the corolla has a trumpet-shaped tube some 3 inches long, ending in a large, reddish-purple, wrinkled and lobed mouth on the lower side of which lie the 4 stamens and pistil with flattened stigma. The outside of the curved tubular portion is white with reddish lines, and the shape and length of this tube decides the species. There is an unpleasant odour to the flower. The corolla falls entire with the stamens attached.

The Fruit is like an elongated marrow, 12-20 inches long and 3-4 inches broad, greyish in colour with a slightly rough skin. The apex is pointed. Scattered about in the fibrous pulp are a number of small seeds which are let loose by the rotting of the fruit, which takes a long time to mature.

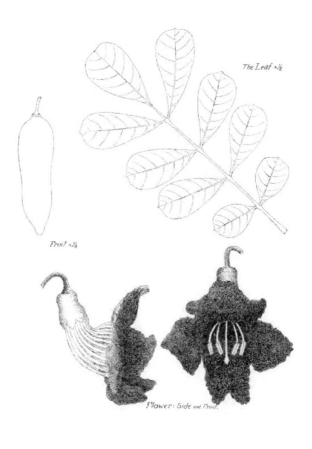

The Leaf ×½

Fruit ×¼

Flower: Side and Front.

LONCHOCARPUS GRIFFONIANUS Dunn.—*Tuburku.*
LEGUMINOSAE.

This is a large deciduous tree which extends up to 11° by way of the streams and "kurmis" and which, below that latitude will grow away, though not far, from their influence. It attains a height of 60 feet with a girth of 8 feet or more and has an enormous spreading crown with drooping twigs and delicate pinnate foliage which gives heavy shade. Bole lengths of 30 feet are seen in close forest and in the open there may be a number of stems from the ground level or near it. The pods, like strips of tanned leather, are the most distinctive feature, as they persist for some time.

The Bark is smooth, with small, very thin, close-fitting scales which remain attached by their centres before falling. The colour is grey or light brown. The slash is pale yellow, soft and thick.

The Wood is whitish-grey. In transverse section the rings are indistinct lines, the pores are very irregular festoons, dense on some zones and single and scattered in the rings. The rays are very fine and close and much waved, and are visible as a fine stippling in vertical sections. In vertical section the pores are open and their brown colour adds a tint to the wood. It is soft, very easy to work with all tools, the planed surface having a slight sheen. Weight 40 lbs. a cubic foot.

The Leaves are pinnate, 9-10 inches long with three to four pairs and a terminal leaflet. The main stalk is very slender and those of the leaflets very short. The leaflets are obovate, with a tongued tip, the lowest pair rounded, the intermediate pairs narrower and the terminal larger and broader. The upper surface is dark and shiny, the under side paler and smooth. The mid-rib is prominent beneath and slightly sunken above. There is a marked contrast between the colour of the leaflet stalks and the under side of the leaflets when viewed from beneath.

The Flowers appear on the leafless tree in February or March. They are on 6-7 inch long spikes all over the bare twigs, and are a delicate blue or mauve colour and pleasantly scented. They are "pea-flowers," ¾ inch long, the calyx reddish-purple, the standard petal with a cream centre splash and the keel white. The flowering period is very short and the leaves follow and cover the tree in a few days.

The Fruits are flat pods, 3-5 inches long and ¾ inch broad, very slightly embossed at the seeds, light brown with a grained surface, very like tanned leather, containing up to 3 flat, bean-shaped, dark brown seeds. A large number of pods fall entire without producing seeds.

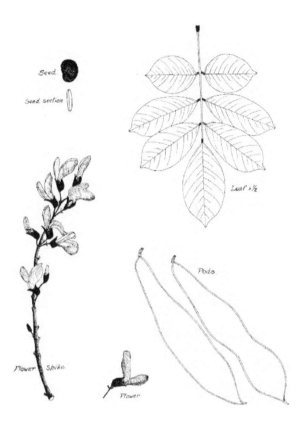

Seed

Seed section

Leaf × ½

Pods

Flower Spike

Flower

LONCHOCARPUS LAXIFLORUS G. & P.—*Farin Sansame, Shunin Biri, Halshen Sa.* LEGUMINOSAE.

A small tree some 15-20 feet high, occasionally 30 feet, occurring commonly throughout the drier savannahs, often in small patches or clumps. It resembles *Stereospermum Kunthianum* (Sansame) in form and leaf but in no other detail. The form is stunted and similar to the majority of the trees in the northern bush savannah, though larger examples with big oval crowns and a 10 foot bole are not uncommon. A curious branching habit is noticed where the main stem grows partly round the base of the side limb forming an enlarged joint. The branches are markedly drooping. The pale bark, purple flowers, narrow flat pods and greyish-green leaves are distinctive features.

The Bark is light grey or yellowish and fairly smooth, the scales being small, close together and corky. Narrow ridges of cork form on the branches and that of the stem thickens for protection against fires. The slash is yellow with black streaks, very distinctive.

The Wood is light yellow, of uniform colour. In transverse section the rings are marked and the concentric lines of hard and soft tissue very plainly seen. The pores are large and small, few, scattered about in the lines of soft tissue in festoons. The rays are straight, variable in width, some visible, others not, to the unaided eye. The areas of hard tissue are clearly divided into rectangles by the sharp lines of soft tissue and the rays. The wood is hard, heavy, not difficult to saw and planing to a smooth finish with a slight sheen. The weight is 55 lbs. a cubic foot.

The Leaves are a foot long with 5-7 leaflets, the terminal longest, the lowest pair shortest. They are grey-green and waxy to the touch and with the exception of the mid-rib the venation is raised on the upper and not on the lower surface. They are 3-4 inches long and tapering to both ends.

The Flowers are in numerous panicles at the twig ends, a mass of purple at first erect, then drooping. The panicles are up to a foot long and the flowers are papilionaceous with a white splash on the face of the standard petal. The calyx is dark purple. They appear from January to March and are fertilised by bees, flies and wasps and other insects.

The Fruits are pods, from 2-5 inches long according to the number of seeds, generally 2, rarely 3, and an inch in width. They are flat, light brown or whitish in colour, in pendulous masses, very conspicuous and the seeds are chestnut brown, ½ inch long, shiny, flat and with a deeply indented hilum. They remain on the tree for a long time.

Uses.—An infusion of the roots with water is used as a tonic.

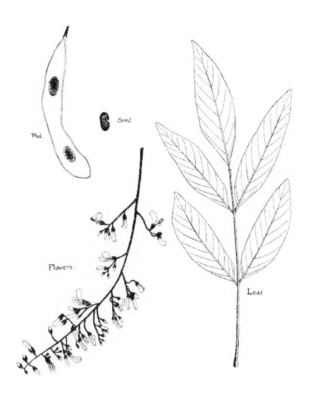

Pod. Seed. Flowers. Leaf

LOPHIRA ALATA Banks.—*Namijin Kadai, Mijin Kadai. "Meni Oil Tree."* DIPTEROCARPACEAE.

This is one of the type species of the tree savannah and strongly resembles at first sight the Shea, since both have strap-shaped leaves. This is the only close resemblance and even they are easily distinguished from one another, see below. It attains a height of some 40 feet with a girth up to 6 feet, and has a tall, regular, rather open crown occupying at least two-thirds of the height and often more. The branches ascend at an acute angle to the stem so that the form is narrowed, another distinguishing feature between it and the Shea. It prefers good deep soils and its northern limit is about 12° N.

The Bark is light grey with even-sized scales 1-2 inches long, 1 inch wide and ¼ inch thick. That of the smaller branches is softly corky. The scales are not prominent, nor is there a milky sap, two other features distinguishing it from the Shea. The slash is crimson with bright yellow edges.

The Wood is a dull reddish-brown. The transverse section shows indistinct light rings, close together, the pores small and scattered singly and widely throughout the hard and soft tissue which is clearly separated in thin, wavy concentric dark and light lines. The rays are exceedingly fine and invisible to the naked eye. In vertical section the pores show as if filled with some chalky matter and the grain is straight with the hard and soft tissue marked in fine lines, and the rays as numerous small bands in radial section. The wood is hard to saw and plane, picking up in places, but elsewhere finishes smooth. It is tough and strong and weighs 65 lbs. a cubic foot.

The Leaves are strap-shaped, some 12 inches long, 3-4 inches broad, with an inch stalk. The margin is waved, the mid-rib alone prominent on both surfaces. The venation is extremely delicate, visible when held to the light. The new leaves are bright red, the red colour fading from the base upwards. The surface is shining, the texture soft. Clustered round the end 3-4 inches of the twigs, they appear like rosettes. The delicate venation, short stalk and red tips distinguish it from the Shea.

The Flowers are in clusters on long stalks amongst the leaves and resemble Apple blossom. They are white, cup-shaped, 1½ inches across, with 5 notched white petals and a ring of numerous yellow stamens round a stout bifid pistil. The calyx consists of 5 sepals, 3 small and round, 2 larger and red tipped, the latter enlarging to form the wings of the seed. They are scented, visited by bees and appear from November to February.

The Fruits are winged seeds, the wings formed by the enlargement of the 2 sepals. At first red, they become light green and leathery, the larger 4 inches, the smaller 2 inches long. The fibrous-coated seed elongates till it is a pointed cone an inch long, light brown as are the now dry wings. The seed contains a single kernel, loose when ripe, about March. Few natives will allow that this tree has any seeds, even when shown them, a popular obstinacy.

Uses.—The wood is too tough to carpenter satisfactorily, but is suitable for building posts. The seed contains some 45 per cent. of oil, said to be suitable for soap after testing, but this is not extracted locally.

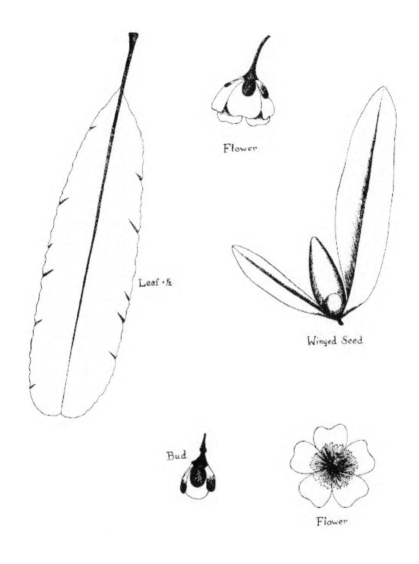

Flower

Leaf ·½

Winged Seed

Bud

Flower

MAERUA ANGOLENSIS DC. —*Chichiwa.* CAPPARIDACEAE.

A small tree, widely distributed throughout savannahs of both types, flourishing equally well in dry sandy soils and richer loamy soils. It is straggling, often growing in the shade of other trees and partly supporting itself by them. In the open it is a slender tree about 15 feet in height, the stem dividing and the open low crown with drooping branches. Occasional examples will reach 25 feet with a 2 feet girth. It extends up to 13° N.

The Bark is dark grey to almost black. That of the smaller branches is dark brown, profusely spotted with small white lenticels which give it a mottled appearance, a most distinctive character. Galls on the new twigs and on the stem are very common.

The Leaves are simple, ovate, but variable in length to breadth, averaging 2 inches long and 1 inch broad. The tip is rounded or slightly cleft with a small projection of the mid-rib. The colour is pale and the surface smooth. The mid-rib is prominent beneath, otherwise the venation is practically unnoticeable. The stalks are ½-1 inch long and the last ¼ inch is thickened and bent at an angle and when the leaf is plucked the stalk nearly always parts at this bend.

The Flowers can be found from November to April and the individual flowers for a long time, almost ripe fruits and new flowers occurring at the same time. They are in leafy racemes of large size. Each has a slender stalk, a long column-like torus, 4 sepals which bend right back and finally fall off, no petals and a number of long white stamens, in the middle of which is the long slender style with the ovary at the tip. The lower part of the calyx is tubular, with the torus inside, and round its mouth is a toothed ring.

The Fruit forms at the end of the pistil, the sepals having fallen, the calyx tube shrunk and the toothed disc being exposed. It enlarges up to 6 inches in length and is constricted at intervals so that it looks like a string of irregular-sized beads. Sometimes two seeds lie alongside, and there may be as many as 30. The fruit is yellow and finally brown when ripe, and is similar in appearance to that of *Cassia goratensis*, though the restrictions are not so narrow as those of that species.

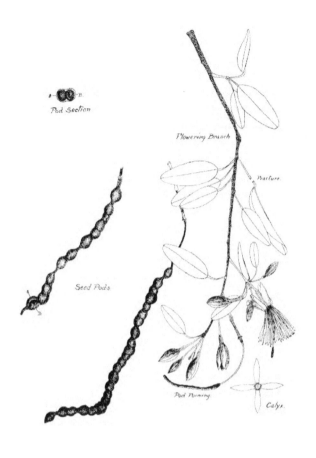

a B.
Pod Section

Flowering Branch

Fracture

Seed Pods

Pod Forming

Calyx

MAERUA CRASSIFOLIA Forsk.—*Jiga*. CAPPARIDACEAE.

A medium-sized tree from 15-30 feet high and up to 4 feet in girth, which grows in dry localities as far north as 14°. It is quite common in N.E. Sokoto. It has a short bole, up to 12 feet, with ascending branches and can be at once distinguished by its heavily pruned appearance, the lateral branches being very short and resembling *Randia nilotica*, the rosettes of leaves being borne on stunted shoots on long, straight, tough branches. The form varies from the short tree with wide, compact crown to the tall, narrow growth of larger specimens.

The Bark is dull grey, smooth in young trees and longitudinally scaled in older trees.

The Leaves are in tufts at the end of short woody shoots and are ¾ inch long and ⅓ inch wide, tapering at the base, with a broad tip, in the shallow cleft of which projects the mid-rib. They are a dull, dark green and limp and leathery in texture.

The Flowers which appear in February rise from the leaf-bearing shoots on ½ inch stalks. Each has a calyx of 4 sepals, light green, which fall back to the stalk with maturity, some 30 white stamens with green anthers, all rising from a short column, and a long pistil with thickened stigma. There is a faint perfume.

The Fruits are like those of *Maerua angolensis*; long, jointed pods formed by the growth of the clubbed pistil. They are brown when ripe and break into sections. The seeds are 10-20 in number and ripen about April.

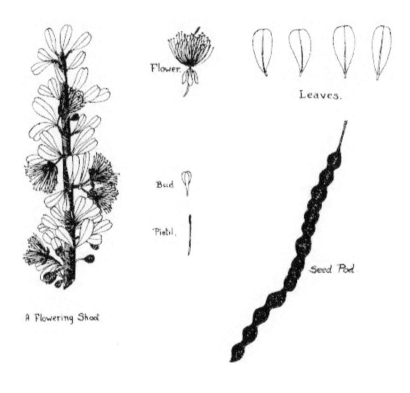

Flower.

Leaves.

Bud

Pistil.

Seed Pod.

A Flowering Shoot

MIMOSA ASPERATA Linn.—*Gumbi, Kaidaji, Kardaji.* LEGUMINOSAE.

A very common shrub which inhabits marshes, the banks of rivers and the fringes of lakes, in fact, any low-lying, inundated ground, forming dense, impenetrable thickets, often of large extent. On the banks of rivers it is frequently associated with a species of *Salix* or willow. It has numerous stems and long erect or drooping shoots, 10 to 12 feet in length. The whole plant is covered with hairs.

The Thorns on the stems are in three's, below each leaf, one thorn directly under the leaf, with a pair a little higher up on either side. The thorns on the stems are all the same size and shape, but those on the leaves differ as described below.

The Leaves are 3-5 inches long with 6-8 pairs of pinnae, bearing up to 40 pairs of long, narrow leaflets. The leaf is sensitive and folds up tight in the evening, the leaflets falling back under the mid-rib. On the upper side, between each pair of leaflets, is a short, slender and very sharp thorn, pointed well forward, and between the lower 4 or 5 pairs is a pair of sharp thorns with broad bases in the same plane as the leaflets. These thorns are absent in the upper part of the leaf. The young leaf is hairy.

The Flowers are in spherical heads about ½ inch in diameter, either white or mauve. Each minute flower has a calyx of 4 sepals, 8 stamens with white or mauve filaments and white anthers, and a long pistil.

The Fruits are pods, 2-2½ inches long, ½ inch wide and ⅛ inch thick, slightly curved. They are covered with erect, bristly hairs whose tips curve forward. On the green pod these hairs are golden brown; on the ripe pod they are unpleasantly penetrating. The pod splits transversely into 15-20 segments, each containing a small bean. The outer rim of the pod remains intact, the sections dropping out in the same manner as those of *Entada sudanica*.

Uses.—It is used as a fence round farms for protection against grazing, and in places where there is an acute shortage of fuel demands a price in the markets.

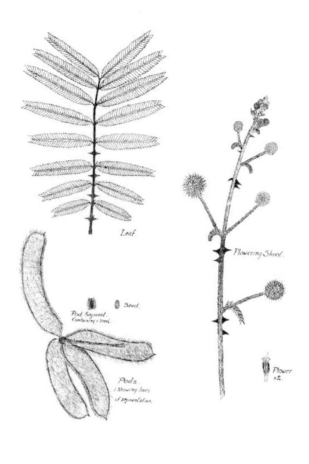

Leaf.

Flowering Shoot.

Seed.

Pod Segment.
Containing 1 Seed.

Pods.
1 Showing lines
of segmentation.

Flower
×2.

MITRAGYNE AFRICANA Korth.—*Giyeya. Giyaiya.* RUBIACEAE.

A large tree which inhabits "fadammas," "tabkis" or annually inundated areas. It has several erect, straight stems ascending to a height of 60 feet in large specimens and the foliage reaches almost to ground level. From the branches, especially noticeable when one of these is horizontally inclined, are a number of slender erect shoots, like those of pollarded willow. The crown is the same width throughout with a narrowed or pointed top. The tree coppices very well and would provide a rotation crop if properly worked. It will not stand drought and a succession of dry years will kill the tree, subsequent wet years only serving to rot the dead trees. This is of frequent occurrence in the farthest north.

The Bark is light brown or grey and smooth, a few scales showing on the lower part of the stem in season. The slash is light brown, rapidly darkening.

The Wood is a very light brown. In transverse section the rings show broad bands of a darker colour, a purple tinge which is clearly seen in vertical, especially radial sections. The pores are minute, quite invisible to the naked eye, and with difficulty seen with the lens, in rows between the extremely fine rays which are waved and very close together. The grain is very close and fine; seasons well and is slightly liable to borer beetle attacks, works well with all tools, the plane producing a nice clean finish of a velvety nature. It is one of the best all-round timbers for many purposes and its small sizes are a great misfortune. The weight is 40 lbs. a cubic foot.

The Leaves vary somewhat in shape from a narrow oval to a broad, rather square oval, both shapes with a pointed tip. They are 3-4 inches long and 2-2½ inches broad with ½-¾ inch stalk. They are in opposite pairs as is so typically seen at the base of the flower or seed head. They are paler beneath than below, very soft and thin in consistency and the venation appears almost white by contrast.

The Flowers are in yellowish balls an inch in diameter, sweet-scented and appearing at the beginning of the rains, though a few blooms may be found at odd seasons. Each flower is about ½ inch long with 5 long sepals narrow at the base, broad at the tip (obconical); a tubular corolla with 5 lobes between each of which protrudes the anther of a stamen, grey in colour; and a long pistil terminating in a reddish stigma, shaped like a bee-hive. The flowers are at first white and then turn yellow.

The Fruits are in spherical heads ¾ inch in diameter and dark brown in colour. Each fruit is what is called di-coccous, *i.e.*, the fruit is formed of two separating carpels, containing seeds. The fruiting heads are most conspicuous and enable the tree to be identified from some distance.

Uses.—The timber has been used lately by the Industrial Schools with great success, and in Sokoto it has been found one of the best for all purposes where large sizes are not necessary. The native makes bowls and spoons from it and it is a good source of poles for building houses. It is supposed to rot rapidly if exposed to wet.

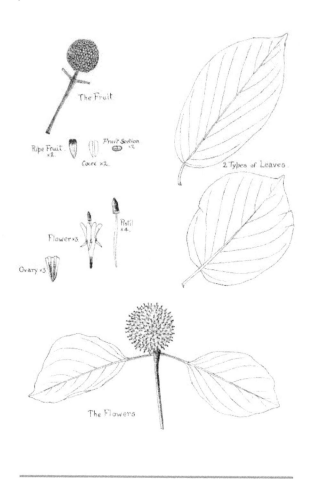

The Fruit

Ripe Fruit. ×2.

Core ×2.

Fruit Section ×2.

2. Types of Leaves

Flower ×3

Pistil ×4.

Ovary ×3

The Flowers

MONOTES KERSTINGII Gilg.—*Wasani, Gasa kura, Farin Rura.* DIPTEROCARPACEAE.

A small tree of the better types of savannah not extending much north of 11°, and very locally distributed. In certain localities, *e.g.,* S.E. Sokoto and W. Bauchi, it is very abundant and over hundreds of square miles will be found pure in patches and elsewhere composing 40-50 per cent. of the growth. It is slender, erect, up to 30 feet in height, with a stem of 1½-2 feet in girth, with the stem bare of branches up to 10 feet and the crown high, open and narrow. Trees in the open will grow large, round, compact crowns. The distinguishing features are the very light under surface of the leaf and the enlarged sepals which surround the fruit. The species resembles *Parinarium.*

The Bark is grey with light patches and fairly smooth, with small, close-fitting, not very prominent scales. The slash is dark red.

The Wood is a light brown-pink. In transverse section the rings are distinct as numerous very close fine lines, the pores are minute, numerous and crowded in small festoons interspersed with numerous single pores, the soft tissue being poorly developed. The rays are so fine and close as to be only just visible with the lens. The grain is very close and wavy and the planed surface reflects the light so as to produce a "shot" appearance, with a sheen. The wood is very hard to work with saw and plane and very cross grained. The weight is 64 lbs. a cubic foot.

The Leaves are oval, some 4 inches long and 3 inches broad, slightly cordate at the base, broad at the tip, with a ½ inch long downy stalk. The upper surface is dark green, shining and rough to the touch, with the venation below the surface, and the mid-rib bends at each lateral. The under surface is grey and velvety, the whole venation being prominently marked. At the base of the leaf there is a curious oval space where the blade of the leaf meets over the stalk at the bottom and again at the third lateral, leaving a gap through which the base of the mid-rib seems to appear.

The Flowers, which appear in the rains, are in small axillary clusters, each about ½ inch in diameter, with a 5-pointed calyx, 5 long white pointed petals, a mass of long stamens filling the corolla and a short straight pistil. They are very inconspicuous.

The Fruits are capsules, ½ inch in diameter, roughly round and wrinkled when ripe, with 3 cells in each of which is a small seed loose in the cell. The capsule is hard, fibrous and loosely porous in transverse section. Round it are the 5 greatly enlarged sepals, modified into wings, bright red at first, then drying pale brown. The fruits are very persistent on the tree, many not falling till the new leaves appear. The red wings are conspicuous from a great distance.

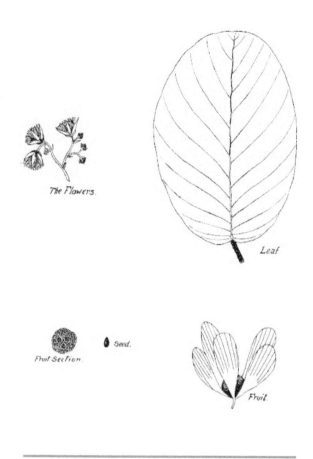

The Flowers.

Leaf.

Fruit Section.

Seed.

Fruit.

MORINGA PTERYGOSPERMA Gaertn.—*Zogalagandi, Bagaruwar Makka.* MORINGACEAE.

This is a very familiar little tree, an exotic from India, which is planted in compounds in the north. It grows some 15-25 feet high with a single stem often forking or dividing into 2 or 3 stems at or near the ground. It is distinguished by its tripinnate leaves, white flowers and long triangular, pointed pods.

The Bark is a deep bluish grey and smooth.

The Wood is white, very soft and of no use whatever.

The Leaves are tripinnate, about a foot long, with some 6 pinnae, each with some 2-5 pinnules, some again with one or two pairs of leaflets and a terminal leaflet. The terminal leaflets are generally larger than the others and all are variable in size and shape from small oval to large spear-shaped leaflets. The foliage is very sparse and graceful.

The Flowers are white, in loose panicles, in flower for many months in the year, during the dry season. Each has a 5-parted calyx with unequal sepals, 5 white petals, the 2 upper ones smaller than the others, 5 stamens with and 5 stamens without anthers, and a slender style. The flowers are sweet-scented.

The Fruits are long, triangular capsules up to 18 inches long, slightly jointed, with 2 grooves down each side, a sharp beak at the tip, light brown in colour and splitting down the 3 edges to release the seeds. These are blackish with 3 papery wings so shaped as to overlap and fit into one another, the bottom of the one seed touching the top of the wings of the one below it. The white pith, with a sheen, is indented to hold the seeds.

Uses.—It is commonly used as a fencing to compounds, the mats or cornstalks of which the fence is composed being tied to the growing trees. The roots are eaten as horse-radish and the leaves fresh as a vegetable. The seeds yield a very fine oil (oil of Ben) which has been pronounced as suitable for watch lubrication.

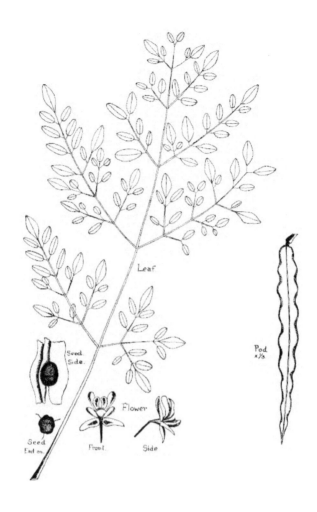

Leaf

Seed. Side.

Seed End on.

Flower

Front.

Side

Pod x⅓

OCHNA HILLII Hutch. OCHNACEAE.

This is a small tree, common locally in parts of Zaria, Bauchi and S. Sokoto, not extending north of 12°. It is as a rule slender and erect, up to 15 feet in height, older trees forming a round crown, younger ones often shrub-like. The yellow flowers and red and black fruiting stage are very conspicuous and the serrate obovate leaves are typical.

The Bark is smooth, a dull grey or brown, with small scales. The slash is yellow, rapidly darkening on exposure.

The Leaves are obovate, 3-4 inches long, 1½-2 inches broad, with cleft tip and finely serrate edges. The venation is prominent on the upper surface.

The young leaves are shiny and tinted red, and the mature leaf is tougher and grey-green with a waxy bloom.

The Flowers are borne on a short woody shoot, 4 or 5 together with 1-2 inch, slender stalks. Each has a calyx of 5 pale green overlapping sepals, 5 yellow petals smaller than the sepals and a ring of numerous stamens round an erect pistil. The flowers appear in February or March and the petals sometimes only remain 24 hours on the flower.

The Fruits are drupes, or small plums, in a ring round the much swollen torus or disc. The sepals enlarge and turn crimson, as does the disc, and the seeds, at first green, turn black and shiny, shrivelling when the surrounding pulp dries up. The branches are often borne down by the weight of the large crop of fruit which is a most conspicuous feature. The elongated pistil and the remains of the stamens appear like bristles on the torus. Most of the seeds are destroyed by a grub.

Seeding Shoot.

Calyx and Seeds.   Calyx. Seeds removed.

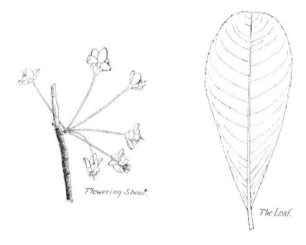

Flowering Shoot.

The Leaf.

ODINA ACIDA Walp.—*Farun mutane.* ANACARDIACEAE.

A large tree, similar in flowers, fruits and leaves to *O. Barteri*, but differing in the absence of hairs on leaves and flower spikes to the extent of those on *O. Barteri*, and distinguished from it at once by the very smooth bark. It reaches large sizes, but not great heights, the average proportion being some 30 feet high with a girth of 6-8 feet. A very short, stout bole divides into a number of large, widely spreading limbs, the twigs drooping down to near the ground. The crown is open and rounded and of great width, affording little shade. Its habitat is the granite country, where it grows in large quantities on the plains or up in the rocky hills, stunted specimens flourishing on almost bare rock with the roots in the smallest cracks. The branches are very flexible and can be bent almost double without snapping. It is called "Farun mutane" as against the "Farun doya" of *O. Barteri*, because it is used for food and medicine.

The Bark is silvery grey, smooth to the point of shining, with occasional scales and a few lenticels. Very old trees cast large scales from the lower trunk. The bark frequently shows a marked spiral twisting round the stem and there are folds and creases like a skin. Resin exudes from the slash which is light red and of crumbling texture.

The Wood is practically the same as that of *O. Barteri*, q.v., for the most part consisting of a soft, dirty white wood, readily sawn and planed and of little value, the small heartwood being red-brown. The weight is about 25 lbs. a cubic foot.

The Leaves are pinnate, 18 inches long with some 4 pairs of leaflets and a terminal leaflet. These are broad at the base, narrowing towards the tip, the margins sinuous, the basal lobes unequal, the short stalks flattened, the upper surface light green, shining, with a few hairs and a pronounced stickiness to the touch. They appear after the flowers. Young leaves are very slightly tinged red.

The Flowers. The spikes of both sexes appear from January to April. The male spikes are up to 9 inches in length, the flowers in irregular clusters with gaps, especially on the lower part of the spike, without flowers. Each has a small, 4-part calyx, 4 petals and 8 short, erect stamens. A few hairs are scattered about the spike, which is scented. The female spikes are some 3 inches long and sparsely flowered. Each has a small, 4-part calyx, 4 petals, the rudiments of 8 stamens and a rounded ovary with 4 stigmas. There is no scent. Though the spikes of both sexes appear well before the leaves, they may persist till the leaves are practically full grown.

The Fruits are similar to those of *O. Barteri* but very shiny. They hang in dense clusters of spikes, ripening from green to red to a purplish black, resembling cherries. There is a thin skin and juicy flesh round a large, hard stone containing a white kernel attached to the top and not filling the cavity. The fruits ripen in June and the crop is often prolific.

Uses.—The fruits are eaten fresh. The bark is boiled in water and the infusion taken as a medicine for stomach troubles.

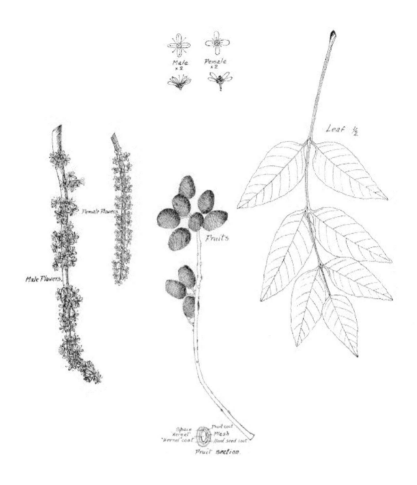

ODINA BARTERI Oliv.—*Farun doya.* ANACARDIACEAE.

A very common tree in open savannahs, flourishing in deep or shallow soils, preferably of granite composition, and plentiful amongst rocks. It is a large tree up to 40 feet in height with girths of 6-8 feet and its characters are a short bole, rarely over 10 feet, and large spreading limbs forming a wide open crown. The branches are very pliant and the twigs droop. This and *O. acida* occur over large areas together, the rough bark and hairy leaves distinguishing the former.

The Bark is almost black and very rough, with deep, vertical fissures and long heavy scales which fall in large ragged lengths. The stem is thus often very shaggy. The slash is salmon pink with paler streaks.

The Wood is dirty white, with bluish and brown discolorations. In transverse section the rings are indistinct; the pores are very small, numerous, single and fairly evenly distributed; the rays fine and close together, hardly visible to the unaided eye. In vertical section the grain is rather coarse, there are no bands of colour and often a number of small knots. The wood is soft, of poor quality, easily worked and of the quality of *Bombax*. The weight is only 25 lbs. a cubic foot.

The Leaves are some 18 inches long, pinnate, with some 4 pairs and a terminal leaflet. These are spear-shaped, about 4 inches long and 2 inches wide, with very short, thick stalks. They are dark green, densely covered with short hairs and limp and flaccid. The young leaf shoots are a reddish-brown and densely pubescent.

The Flowers are in spikes, the male and female on separate trees. The male are in 5-6 inch spikes, up to a dozen spikes at a twig tip, often so numerous on the leafless tree as to render it conspicuous from a great distance. They appear from January to April, and are scented. The flowers are in clusters round the spike, not completely covering it. Each has a 4-lobed calyx, 4 small petals, 8 erect stamens and the rudiments of an ovary with 4 stigmas. The flowers fall from the spikes, leaving them bare. The female flowers are on shorter spikes, the flowers are fewer in number and evenly distributed on the spike. Each has a 4-lobed calyx, 4 small petals and an ovary with 4 stigmas. They are not scented. Both spikes are very pubescent.

The Fruits are nuts, flattened ovoids with the stigmas prominent at the tip. They ripen about March to May and are reddish purple, the flesh very resinous and the white kernel attached to the top. They are ½ inch long and are borne in heavy clusters of spikes, generally erect and stiffer than those of *O. acida*.

Uses.—No part of the tree is edible as in the case of *O. acida*.

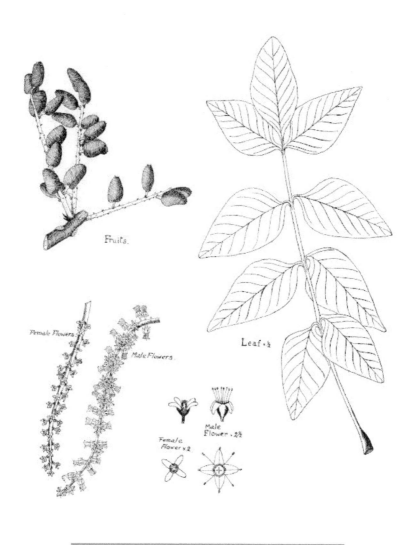

Fruits.

Female Flowers

Male Flowers

Leaf × ½

Male
Flower × 2½

Female
Flower × 2

ORMOCARPUM BIBRACTEATUM Baker.—*Faskara giwa. Tsa.*
LEGUMINOSAE.

A small, erect tree up to some 25 feet in height with 1-2 feet girth, locally very plentiful but otherwise infrequent, *e.g.*, in the Anka district of Sokoto it occurs in vast numbers, but nowhere else in the province, and in one or two places in Bauchi small clumps of it are seen. At first sight it is very like an *Acacia*, with its erect branches and pinnate leaves. The stem branches generally low down, occasionally at a height of 8-10 feet, and the branches are vertical, so that the tree has no width of crown. Owing to the scarcity of the smaller branches the vertical ones are crowded with leaves, and from

February to April with flowers, for several feet, there being no side twigs, these being replaced by small shoots covered with bracts.

The Bark is silvery grey with a soft sheen, smooth and very thick, large scales peeling off in season. The effect of fire on young stems is to produce great bosses of light brown cork, often charred. The slash is yellow.

The Leaves are about 2-3 inches long, pinnate, with some 6-8 opposite or sub-opposite leaflets and a terminal leaflet. These are ¼-⅜ inch long and about ³⁄₁₆ inch broad, oval, with or without a slightly notched tip in which the mid-rib protrudes slightly. They are grey-green and hairy and each springs from between a pair of bracts on the twig.

The Flowers are very handsome "Pea-flowers" in small clusters on long stalks, branched, with a bract at each branch, springing from small clusters of bracts on the woody stems. The stalks are purple and hairy, the calyx is irregular, purple and green, and the corolla is a delicate pink with a pale green keel. The flowers are generally in great masses and a curious characteristic is that they dry up and retain almost their original shape, the pod ripening meanwhile.

The Fruits are small black jointed pods 1-1½ inches long with some 4-5 joints, oval, flattened and hairy, each containing one seed, pale brown in colour. The pod will often bend right round and its tip enter the mouth of the flower.

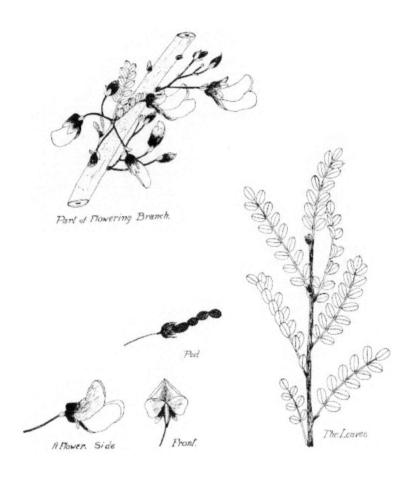

Part of Flowering Branch.

Pod

A Flower. Side

Front.

The Leaves

OSTRYODERRIS CHEVALIERI Dunn.—*Durbi.* LEGUMINOSAE.

A large tree some 40 feet or more high, common in some parts of Sokoto, Zaria and Katsina, but very local in its occurrence and extending as far north as 13°. The girth is up to 7 feet, more often 5 feet. It closely resembles *Paradaniellia Oliveri* in bark and leaf and form, especially in young trees of the latter species. The bole may be clean and straight for 20 feet and the crown is high and flat-topped in full grown trees. It does not grow on poor soils and its extension to 13° is by way of moister valleys and pockets of alluvial soil. The pods and flowers are the distinctive features.

The Bark is light grey with a strong resemblance to that of *Paradaniellia Oliveri*, the scales being even-sized, polygonal but not so large as those of *P. Oliveri*. Nor has it the red tinge of the latter. The slash is most distinctive, being a mixture of fibres of white, dark brown and crimson, like strands of wire.

The Wood is whitish or cream-coloured. In transverse section the rings are indistinct but the concentric lines of hard and soft tissue are well marked. The pores are single or in small groups, unequally scattered. The rays are invisible to the unaided eye, and waved slightly. In vertical section the pores are long and open, the hard and soft tissue well marked in parallel lines and the rays show as long, fine bands on the radial section. The wood is soft, medium weight, shows red streaks, is easily worked and the planed surface is smooth. It has a curious smell, The weight is 45 lbs. a cubic foot.

The Leaves are pinnate, 10-15 inches long with some 6 pairs and a terminal leaflet. The leaflets are opposite or nearly so and increase in size towards the top pair, the largest 3 inches long and 1½ broad, with cordate base, unequal lobes and an even taper to a rounded, cleft tip. The upper side is bluish-green with a bloom, the underneath pale grey-green. The venation is prominent beneath.

The Flowers, from February to April, are in erect panicles on the stout twig ends, some dozen or so on each twig. Each flower is white, papilionaceous, ¾ inch long. The calyx is covered with brown hairs, as are the flower-stalks and there is a green splash in the centre of the standard petal. The panicles are 6-8 inches long, gradually drooping.

The Fruits are pods, at first limp and green, then brittle and dark brown. They are 3-6 inches long, 1¾ inches broad, flat and strap-shaped, the surface veined, and a prominent ridge runs round the outside a little way from the edge, outside which ridge there is no veining. The seeds, more often 1 than 2, are slightly embossed, and are flat, brown, round beans with a white hilum. The pods ripen from May onwards and numbers of them persist on the tree till the following flowering season.

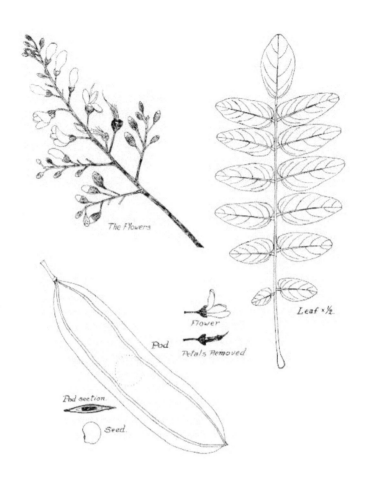

The Flowers

Flower

Pod

Petals Removed

Leaf × ½.

Pod section

Seed.

PARADANIELLIA OLIVERI Rolfe.—*Maje, Kadaura.* *"Copaiba Balsam Tree."* LEGUMINOSAE.

This is one of the largest trees of the savannahs and gives an enormous volume of timber. It reaches a height of 80 feet and girths of 12 feet are common. A girth of 26 feet combined with a 40 feet bole has been measured. It has a gregarious habit similar to *Khaya senegalensis* on the edges of "fadammas" and in flat country where water lies during the rainy season. Hundreds of saplings may be found round these clumps. It is most readily distinguished from other species, at a distance, by the straight, light-grey stem, which is generally tapering in form, and by the shape of its crown which is triangular with a flat top. The crown is dense and dark, the limbs at an angle of about 30 degrees to the bole and the bulk of the foliage being on the summit. The bole is generally swollen at the base. It is a very fast growing species.

The Bark is very light in colour, this being emphasised at a distance. It is rough with large even-sized scales. The bark of saplings, owing to the annual fires, is very thick and scaly, cracking in horizontal rings round the stem. This forms a highly protective covering. An oleo-resin, or balsam, exudes from the slash, which is dark crimson with small white streaks.

The Wood, of which a great volume is yielded by the tree, is red-brown with darker streaks. In transverse section the rings are fairly well defined dark, regular lines, the pores are large, mostly single, widely and fairly evenly distributed, the rays fine, continuous, unevenly spaced but 3 to 4 to each pore, bending in a curious manner where they cross the lines of soft tissue, just visible with the naked eye. In radial section the rays are small light-reflecting bands and in the tangential section are clearly visible as fine ripples. The sapwood is whitish with a faint brown or pinkish tinge, and the rays are very clearly marked on the two vertical sections. This is a very good light-weight timber, very easily worked and not very durable or strong, but its large sizes, soundness and appearance enhance its value. It has a most pleasant cedar-wood smell. Weight 44 lbs. a cubic foot.

The Leaves are 18 inches long, pinnate with some 6 pairs of large, pointed, dark-green, shiny leaflets. The young leaves are soft and pink in colour.

The Flowers are conspicuous, chiefly at the top of the tree, in large, flat panicles, in December. The flowers stand erect on the horizontal panicle and consist of a 5-lobed, imbricate calyx, one sepal being smaller than the others, greenish-white in colour; 10 white stamens, 2 inches long and a long, white pistil which swells to form the seed-pod. The flower parts are borne on a stout club-shaped receptacle.

The Fruit is a flat pod, about 3 inches long and 2 inches wide, whitish in colour. It splits by the curling up of the inner layers and frees one oval, flat, dark-brown seed, the rudiments of 3 or 4 others being visible attached to the suture. The open pods with large brown seeds hanging from their edges are a conspicuous feature in masses on the tree tops, the seeds and pods remaining some time in position before being blown down.

Uses.—The timber is made into planks, mortars, canoes and cattle-troughs. The resin is burnt in houses as a fumigating incense.

In times of extreme famine, the young leaves are appreciated as a vegetable, being eaten with the addition of salt and pepper.

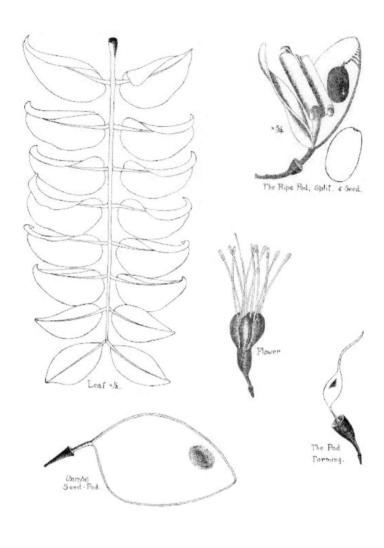

The Ripe Pod, Split. & Seed.

Flower

Leaf ×⅓.

The Pod Forming.

Unripe Seed-Pod.

PARINARIUM CURATELLAEFOLIUM Planch.—*Rura, Gwanja kusa.*
ROSACEAE.

A small and very common tree averaging 15-20 feet high with a girth of 1-2 feet. It has no form or special characteristic of growth, being similar to the many other species with which it occurs in open savannah forests. Its only importance lies in its occurrence in large quantities over considerable areas. It is readily distinguished from other species by its flowers and fruits, which latter may be seen for several months in the year.

The Bark is very dark, sometimes almost black, with small, prominent, corky scales up to 1 inch square. The slash is a dull, dark red.

The Wood is light brown, with a slight orange tinge. In transverse section the rings show as slightly darker lines, the pores are small, oval, with their length radial, single, not numerous, and connected by continuous waved concentric lines of soft tissue. The rays are extremely fine and very close together, quite invisible to the naked eye. In vertical section the pores are open and slightly darker, the grain close. It is a soft, sound wood, easily worked, planing with a dull, smooth finish, not picking up much. The weight is 48 lbs. a cubic foot.

The Leaves are about 5 inches long and 2½ inches broad. They are alternate, pale green and smooth above and greyish beneath.

The Flowers are in long, dense or open, terminal panicles, a number together. They are greenish in colour, the whole panicle covered with a light pubescence. Each flower consists of a grey-green, 5-pointed calyx, 5 minute white petals and 10 stamens, some with pink anthers. They appear about December or January, and are conspicuous by their numbers on the leafless tree.

The Fruits are reddish-brown drupes, about an inch long, the skin covered with a number of small, grey lenticels. They have a reddish, sweet, edible flesh and a large, hard stone. They ripen towards the end of the year and are the readiest means of identification of the species.

Uses.—The fruits are eaten fresh.

The young trees are cut for poles for building (gofa).

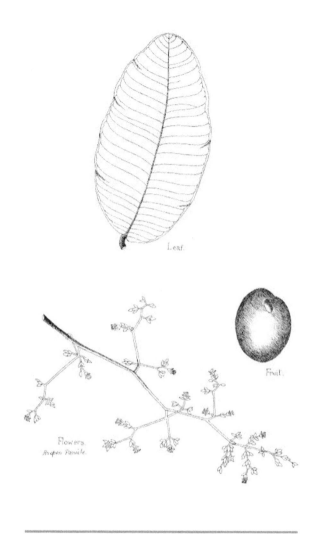

Leaf.

Fruit.

Flowers.
An open Panicle.

PARINARIUM MACROPHYLLUM Sabine.—*Gawasa.* *"Ginger-Bread Plum."* ROSACEAE.

A common tree locally in parts of the more northerly provinces. It is not found distributed evenly through the forests but in numbers over small areas here and there. It is very highly valued by the native for its fresh fruit and protected on farm lands. It is a thick-set, squat tree with a short, heavy stem and wide-spreading, crooked branches forming a round or flat crown. It is from 15-25 feet high, sometimes more, with a girth up to 8 feet. The foliage, owing to the size of the leaves, is dense. It is somewhat similar to its congener, *P. curatellaefolium*, but the leaves are larger, the flowers heavier and the whole tree on a larger scale. It grows in dry, sandy soils.

The Bark is grey and not rough, covered with small even-sized scales.

The Wood is a light brown colour, frequently with large grey discolorations darkening the colour. In transverse section the rings are obscure bands, the pores are small, few and distributed rather unevenly, single, connected by very faintly marked and poorly developed soft tissue lines. The rays are exceedingly fine and very close together. In vertical section the pores are few and fine and the grain is close. It is a fairly hard wood, sawing well and planing with little picking up to a hard, smooth finish which polishes well. Weight 45 lbs. a cubic foot.

The Leaves are some 5 inches long and 3-4 inches wide on the average, but may be as much as 8 inches long and 5 inches wide, with a stout stalk not above ¼ inch long. They are slightly cordate at the base and broad at the tip with a small point. There are some 15-20 straight veins on each side of the mid-rib, sunken on the upper surface and very prominent on the underneath. The upper surface is a brilliant light green gradually darkening with age, the under side grey. There is a dusty brown covering of hairs on the upper surface, which rubs off readily in the hand and exposes the smooth green surface. The stalk is densely covered with brown hairs.

The Flowers are in terminal racemes from 3-9 inches long, all parts except the petals covered with the hairs. Each flower has a 5-lobed calyx, 5 white petals in 2 groups of 3 and 2, 10-20 white stamens with yellow anthers and a curved, white pistil. The flower in bud is enclosed in 2 bracts which are pushed up by expansion of the flower and then fall off the end. The flowers are found on the tree every month of the year, a continual succession of flowers and fruit occurring.

The Fruits, which are found all the year round, are plums, reddish-brown with an orange tint, rounded or oblong, 1½-2 inches long and 1-1½ inches broad, the surface roughened with numerous grey lenticels. The sweet, edible flesh is whitish, moist and mealy, about ½ inch thick round the very tough, thick stone which has one or two "kernels." The ripe fruit has a sickly sweet smell.

Uses.—The wood of large specimens makes very good mortars.

The fruits are highly appreciated and eaten fresh. They are brought to the big towns from long distances in some places and will sell for as much as a penny each in the markets.

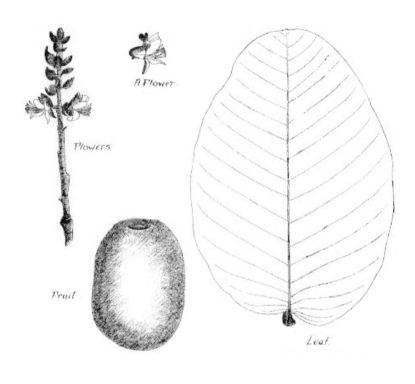

A Flower.

Flowers.

Fruit

Leaf

PARKIA FILICOIDEA Welw.—*Dorowa. "Locust Bean Tree."*
LEGUMINOSAE.

This is the type tree of the Park formations and a very common and well-known species with a wide distribution. Owing to its great value as a food and its uses by the natives, it is often the sole occupant of extensive areas of cultivation. It grows to large sizes, 50-60 feet high with girths of 10 feet or more. The bole, in the case of trees which are grown in the open, is short, not, as a rule, above 10 or 12 feet, and several large, spreading limbs form a very wide crown which appears dark and dense at a distance, but which, in fact, gives only average shade. There are small, rounded root-flanges. All trees near habitation are owned by the individual. Those in the forests have a different habit of growth, the bole being longer, the branches more erect and the crown less spreading. The large crowns of the park trees are generally due to the cutting of the branches for fuel. *See* USES. The tree pollards well and shoots readily from the stool. It is liable to attack from white ants when young, and to the ravages of caterpillars which may completely defoliate the tree.

The Bark is dark brown or dull grey, with small, regular scales of varying roughness. The branches and the stems of young trees are a light grey, almost silvery, and smooth. The bark at the base of the tree is often much chipped about by the natives who make an infusion of it, drunk as a tonic. The slash is brick red, spongy and fibrous in thin layer formation.

The Wood. The heartwood is a dull brown colour, the sapwood a dirty yellow. Even the largest stems, 3-4 feet in diameter, cut in the north, show none of this heartwood. In transverse section the rings are faint and wide apart, the pores are large, open, fairly regularly distributed in the well-marked soft tissue festoons, with nests of 3 or 4 here and there. The rays are straight, continuous, evenly spaced, not all the same thickness, and show as long, light-reflecting bands in radial section. In vertical section the grain is fairly open, the hard and soft tissue giving a mottled appearance in tangential section. The wood is most easily worked with all tools, planes with a dull finish and though liable to crack a little is on the whole a sound, though weak, timber. The weight is only 32 lbs. a cubic foot.

The Leaves are bipinnate, 15-18 inches long, with opposite, pinnate leaflets bearing numerous dark green pinnules, oblique, ¾ inch long, alternate but nearly opposite towards the tip. They have a shining surface, and the young leaf is reddish.

The Flowers are pendulous balls some 2½ inches in diameter on a 7 inch stalk. They appear in March before the new leaves or with them and are dark red in colour and sweet-scented. They consist of 10 long, red stamens with black anthers and a simple style, in a 5-lobed calyx with honey glands. The flowers are visited both by bees and flies. Called "tutu" by the native, boys suck the balls for the honey. The flower-buds are called "gundar tutu."

The Fruit is a bean, some 9 inches long, dark brown in colour, in pendulous clusters on the tree in July. When young they are called "sabada"; when brown, but before ripening, they are called "garda." The ripe seeds are black and embedded in a yellow, mealy pulp. They are of great value as an annual food-crop.

Uses.—The timber is used for making mortars and basins. The seeds are fermented and made into cakes called "daudawa."

The husks are soaked in water till fermentation takes place and the resulting liquid spread over earth floors and on the walls of dye-pits to bind the surface. The liquid is called "makuba" and is also used as a fish poison.

The branches are cut for fuel for the boiling of the water used in the purification of women after child-birth.

The mealy pulp round the seeds (garin dorowa) is used in soup.

An infusion of the bark is a tonic.

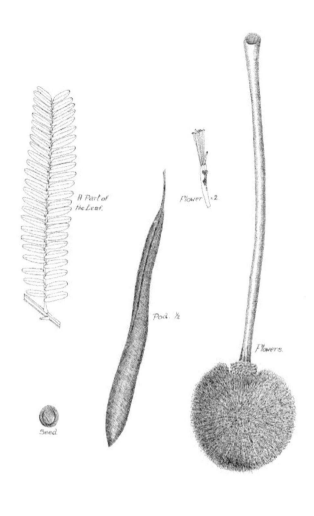

A Part of
the Leaf.

Flower x 2.

Pod. ½

Flowers.

Seed.

PARKINSONIA ACULEATA Linn.—*Sasabani, Sharan labbi. "Jerusalem Thorn."* LEGUMINOSAE.

A small exotic tree introduced by Arabs and planted in most of the Northern towns to such an extent that it is worthy of inclusion here. It is a delicate and graceful shrub-like tree up to 25 feet high. It has a single stem branched a foot or two above ground level; in some cases there is a stem over 6 feet in length before the branches, erect, with their ends spreading out and drooping down, occur. It grows best in an open position and does not thrive in the shade, where it grows very weak and straggling, without the vivid green of trees in the open.

The Bark is bright green and smooth.

The Thorns are straight and sharp, ¼ inch long, light red in colour with thickened, fleshy green bases from which spring the leaves. The scars of fallen leaves can be seen on this thickened base.

The Leaves are bipinnate with one or two pairs of pinnae which consist of a broad, flat mid-rib some 9 inches long from which little oval leaflets under ¼ inch long spring alternately. These are sensitive and at night, or when gathered, fold inwards and lie flat on the mid-rib. The rib and leaflets are dark green and smooth.

The Flowers are in pendulous, sparse spikes amongst the leaves. Each is an inch in diameter and has 5 pale green sepals, 5 crinkly yellow petals, one of which is more prominent than the rest and forms a throat covered with small red spots. There are 10 short stamens with red anthers and a short brown pistil. The flowers are found most months of the year.

The Fruits are jointed pods, varying in length according as they contain 1-6 seeds, the longest being 6-7 inches long. They are jointed, light brown and veined, and when ripe are very light and brittle. The seeds are ⅓ inch long, black, oval, round in cross section, very hard and smooth and rattling loose in the pod. They may be found for some months in the year, particularly in February.

Uses.—Ornamental and shade providing.

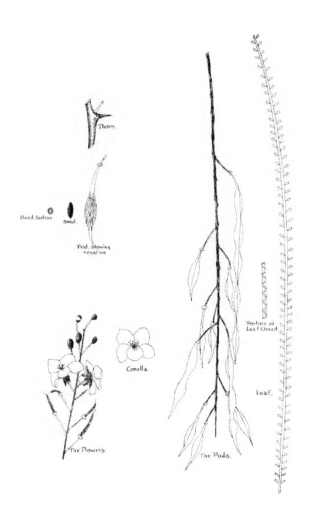

Thorn.

Seed Section. Seed.

Pod, showing venation.

Corolla.

Portion of Leaf Closed.

Leaf.

The Flowers.

The Pods.

PROSOPIS OBLONGA Benth.—*Kiriya.* LEGUMINOSAE.

A large tree attaining a height of over 60 feet with girths of 6-7 feet. In good forest it may have a bole length of 30 feet, but in the open savannahs it is branched to within a few feet of the ground, as a rule, and its crooked limbs form an irregular, rather open crown. The finer twigs droop considerably and the tree may bear some resemblance to *Tamarindus*. In old trees in good forest there is a high, wide crown, open, and giving little shade. There are small rounded root flanges. The features which most readily distinguish it are its bark, colour of foliage and pods, q.v. It likes good, open soils.

The Bark is almost black, with ragged, crisp, curling scales which are concave on the outer surface and leave light brown patches where they fall.

Sometimes the bark has a bluish tint. That of the smaller branches is light grey or brownish and smooth. The slash is reddish, darkening to orange and red-brown.

The Wood is a rich red brown. The sapwood is grey. In transverse section the rings are indistinct, the pores are open, regularly distributed, mostly single, a few double. The rays are very fine, waved and unevenly spaced, not visible to the naked eye. In vertical section the pores are seen to have dark resin contents, the grain is open, and there are bands of colour, the lighter soft wood picking up and the darker hard tissue planing smooth, according to direction of planing. There is a marked reflection of light from these bands. It blunts axes, is hard to saw, must be finished with glass-paper after planing, when a fine surface is obtained, oily to the touch. It will not take nails. A very tough, strong, durable timber. Weight 65 lbs. a cubic foot.

The Leaves are bipinnate with 2-4 opposite pairs of pinnae each bearing some 6-10 leaflets ¾-1 inch long and ¼ inch or more wide, with a pointed tip and the rib not in the middle. They are light green and softly pubescent. The pairs are inclined to the top side of the main stalk and between each pair of pinnae there is a little gland. The base of the stalks is much enlarged.

The Flowers are in short pedunculate spikes, the latter 1½-2½ inches long and densely composed of small yellowish flowers, which appear in May. Each flower has a 5-lobed calyx, 10 stamens with yellow anthers and a short hairy pistil. The flowers are sweet-scented.

The Fruits are large pods from 4-6 inches long and an inch in diameter, slightly flattened in one direction, pointed at the tip, a dark, purple-brown colour. The hard, shiny, brown seeds, oval and ⅓ inch long, are embedded in a dry, cream-coloured, spongy-looking pulp, each separated from the other by a thin, transparent membrane. The seeds rattle loose in the pod. The pods are very persistent on the tree.

Uses.—The natives use the wood for tool handles and burn it for charcoal, the quality of which is very fine and much valued by blacksmiths.

The wood is also used for making pestles for large mortars and the manufacture of tobacco pipes.

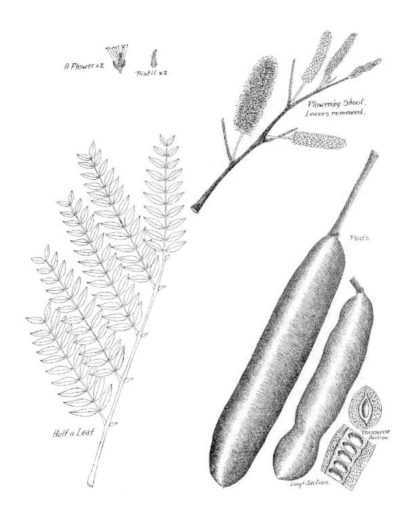

A Flower x2   Pistil x2

Flowering Shoot,
Leaves removed.

Pods.

Half a Leaf

Transverse Section.

Long.^t Section.

PSEUDOCEDRELA KOTSCHYI Harms.—*Tuna(s)*. MELIACEAE.

This species, which is closely allied to the mahoganies (*Khaya*), is a large and valuable timber tree and occurs particularly in groups, or singly here and there in the Tree savannah. It is a handsome tree, regular in shape, with a straight stem and a bole of often 20 feet in length with a girth of 5 or 6 feet. The total height may be 40 feet. It is rather slow growing and though the crown is fairly dense, is essentially light-demanding and not much of a soil-improving species. It can be seen reproducing itself from root-suckers and very rarely from seed, since the latter is destroyed by fire. The crown is cylindrical and regular and the limbs ascend vertically.

The Bark is very conspicuous by its light grey, almost silvery, colour. It is regularly fissured and the scales are of equal size and soft and thick. The base of the stem is often swollen and the bark charred by repeated firing. The slash is bright crimson and shows a layer formation.

The Wood is a red-brown colour, flecked with black and dark brown of a most distinctive pattern, rather open grained, scented, hard and weighs 50 lbs. a cubic foot. It carpenters well and polishes. The pores are small and open, scattered rather thinly along and between the rays which are long and evenly spaced and straight. A useful and ornamental wood if large sizes could be obtained.

The Leaves are pinnate, about 1 foot long with 4 or 5 pairs of leaflets which have wavy edges so that the leaf resembles that of the oak. The leaflets may be opposite or not and a terminal leaflet is not always in evidence. In colour the leaf is dark and shiny on the upper surface and soft and grey with rough venation beneath. They are in a rosette-like formation like the mahoganies.

The Flowers are in panicles in the axils of the end leaves and are white and sweet-scented. Each is about ¼ inch across, with 5 petals, a knobbed pistil and 10 stamens whose filaments are united and surround the pistil. They appear in February.

The Fruits, which ripen in February of the following year, are capsules about 4 inches long, shaped like a club and standing erect. They are brown in colour and split from the apex into five sections which curl back and release the seeds. These are packed, about six in each section, round a pith centre, the wings, 1¼ inches long, pointing downwards, one over the other. The split capsule remains some time on the tree after the seeds have fallen. The capsules are very few in number on a tree considering the enormous number of flowers which bloom, but are very conspicuous and a ready means of identification. The mahogany coloured seeds will blow 200 yards away.

Uses.—An infusion of the bark is used as a cure for digestive trouble. The timber is used for furnishings and is an excellent wood for all articles, owing to its being so easy to work.

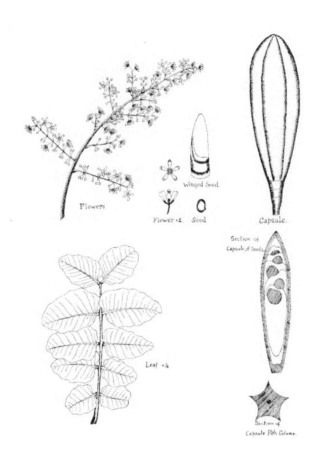

Flowers

Flower ×2   Seed

Winged Seed.

Capsule.

Section of
Capsule & Seeds.

Leaf ×⅓

Section of
Capsule Pith Column.

PSOROSPERMUM SENEGALENSE Spach.—*Kashekaji. Kaskawami.* HYPERICACEAE.

A common shrub in the Bush savannahs, not extending very far north, but abundant in secondary growth after cultivation. It is common amongst the rocks on the Bauchi plateau almost up to the 4,000 feet level. It is equally common all through Zaria and S. Sokoto, in fact, anywhere up to 12° N. It forms a large, compact, round shrub some 15 feet high with dense foliage, generally on a single, low-branched stem. The distinguishing character is the rusty-coloured underside of the hairy leaves.

The Bark is light brown and has small corky ridges and fissures.

The Leaves are from 3-5 inches long and 1½-2½ inches broad, in pairs, ovate, with wavy margins, tapering tips, very short stalks, downy on both surfaces, that on the upper surface rubbing off to expose a light, shiny green. The under surface is densely covered with rust-coloured down, readily rubbed off and exposing a fine network of veins with a minute black dot in almost every cell and a pronounced row of dots all round the margin. The mid-rib is recurved causing the leaf to fold up along it. The shoots are covered with orange-coloured hairs.

The Flowers are in dense clusters 3-4 inches in diameter, from January onwards. Each is ¼ inch across, has a 5-pointed calyx with vertical purple lines on the inside, 5 white petals densely covered with white hairs, 5 columns of stamens with some 9 anthers each and a pistil with 5 lobes each with a globular stigma, shining brown.

The Fruits are capsules with 5 cells, one generally maturing at the expense of the rest. They are a little over ¼ inch in diameter, purple, shiny and fleshy. The sepals clasp them close, and the stigma remains are persistent.

Uses.—A concoction of the leaves and bark is applied for skin diseases.

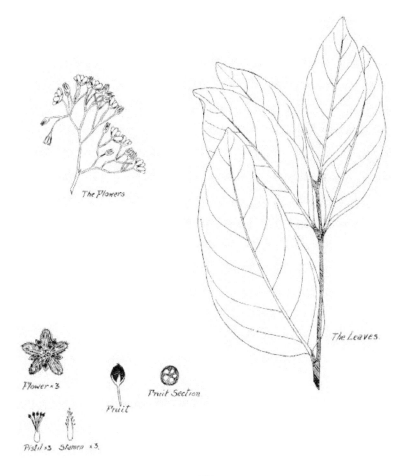

The Flowers

Flower × 3

Fruit

Fruit Section

The Leaves

Pistil × 3 Stamen × 3.

PTEROCARPUS ERINACEUS Poir.—*Madobia*. LEGUMINOSAE.

A medium-sized tree of the savannahs, averaging some 40 feet in height with a girth of 4-6 feet. Timber producing examples considerably larger are common enough and good, clean, straight boles 20 feet or more in length are found. It has a wide distribution up to 14° N., but is a small tree at this latitude. The crown is high, open, wide and rounded, not giving much shade. Larger examples have short buttresses. The species is susceptible to fire which hollows the stems. The long pinnate leaves, masses of yellow flowers and winged, hairy seeds on the leafless tree are characteristic features. Over many miles of the better types of savannah it is the type tree amongst others and occurs in quantities.

The Bark is very dark and rough with scales which curl up at the ends and make the bole appear shaggy. The scales fall in pieces 3-4 inches long. The bark of the branches is light grey and smooth. The slash is brown with fine red lines and a blood-red resin exudes from it.

The Wood is a rose-red or rich brown colour with a wavy grain of darker streaks. It might almost be called figured, especially in tangential section. In transverse section the pores are small, single and scattered and connected by fine, wavy lines of soft tissue; the rays are extremely fine and closely spaced, not visible to the unaided eye, and the rings are indistinctly marked darker lines, generally varying in width apart and rarely concentric. The radial section has a marked banding of dark and light. The wood is hard, liable to split in seasoning, difficult to saw and the planed surface smooth, slightly oily and taking a high polish. The weight is 60 lbs. a cubic foot.

The Leaves are pinnate, a foot long, with some 11-13 pinnae, alternate, and wide apart. The upper surface is shiny, the lower grey-green, the mid-rib prominent beneath and grooved above. The leaves are pendulous.

The Flowers appear from December to February on the leafless tree, and are in such numbers as to be conspicuously yellow from a great distance. The flowering tends to be patchy, that is, part of the tree may be smothered in flowers on one side and not on the other. The flowers are in short spikes with a bright green calyx and a papilionaceous corolla of much-wrinkled yellow petals, which turn pale when fertilised.

The Fruits are winged, round, 2 inches in diameter with the wing much waved and the embossed seed portion covered with short, bristly hairs which lie in all directions. They are often so numerous as to give the tree the appearance of being in leaf. They dry light brown.

Uses.—Larger specimens yield timber suitable for construction and furniture. Smaller trees are used for building posts.

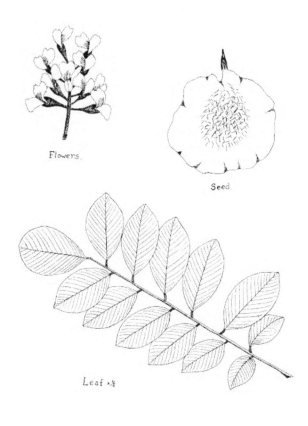

Flowers.

Seed

Leaf ·⅛

RANDIA NILOTICA Stapf.—*Chibra.* RUBIACEAE.

A shrub or small tree, readily recognised by its stiff, spiny stems densely covered with the small leaves. The stems spring from the base and are either erect or finally bent out and down. Occasionally it forms a dense shrub of large size, but usually is light and open. It occurs as far north as 13°, and is common in open high savannah forests.

The Bark is light grey, sometimes almost white, with black scales scattered about in patches.

The Thorns are above the leaves and are short, sharp, very strong and woody, with bark at the base and smooth tips.

The Leaves are about 1½ inches long, narrow at the base, broad at the tip, and smooth; the veins slightly raised on both surfaces. They are borne in rosettes at the ends of very short twigs, above which is a thorn, and densely cover the branches.

The Flowers spring from amongst the rosettes of leaves and may be so numerous as to cover the twigs with a mass of bloom. They are white, changing to yellow and are slightly scented. Each has a short-stalked tubular calyx with 5 unequal-sized sepals with cleft tips, a tubular corolla about ¾ inch in diameter with 5 petals, bent back, 5 stamens in the angles between the petals, consisting of dark brown anthers only, and a yellow, protruding, clubbed pistil. The flowers appear in May.

The Fruits, ripening in June, and often persisting for several months, are oval, 5-8 inches long and ½ inch in diameter, brown, shiny, hard, with thin "skin" vertically ribbed and cellular veined. It is 2-celled and the partition is vertical, each cell containing a number of small black wedge-shaped seeds packed close and cemented together. The fruit falls entire and rots.

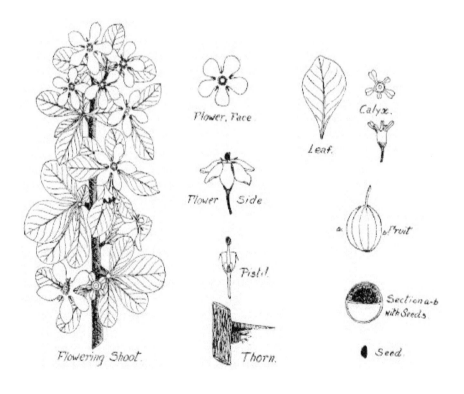

Flowering Shoot.   Flower, Face.   Flower Side   Pistil.   Thorn.   Leaf.   Calyx.   Fruit   Section a-b with Seeds   Seed.

RAPHIA VINIFERA P. Beauv.—*Tukuruwa*. PALMACEAE.

This very common palm grows in swamps, where it forms impenetrable thickets often of large extent. It is readily distinguished by its very long erect leaves growing from ground level, there being no stem. It reaches a height of over 30 feet, the girth of the base being some 8 feet. The leaves spread out from a growing centre, visible as an erect shoot in the middle.

The Leaves are upwards of 30 feet in length with long leaflets whose edges and mid-rib at the back are armed with small sharp spines. The mid-rib is an orange colour, turning grey with age.

The Flowers are in a loose branched spadix up to 6 or 8 feet long, the male and female on the same palm and on the same spadix. The males on the lower part of the branches of the spadix have a tubular calyx, a corolla of 3 petals, long and narrow and a number of stamens, varying from 10-12 inside the base of the petals, attached to them. The female flowers, situated at the end of the branches of the spadix, are larger than the males, the calyx tubular, the corolla bell-shaped with 3 sharp lobes, the rudiments of the stamens forming a cup attached to the corolla and a 3-celled ovary.

The Fruits are like cones, some 3 inches long and 1½-2 inches broad, smooth and shiny, with a number of broad, pointed scales, varying in colour from chestnut to red-brown. They contain one kernel, loose in the shell.

Uses.—The leaflets are used for plaiting into mats, &c. The mid-rib is used for roofing material, beds, canoe poles, and split into lengths, the inner portion makes mats. The sap is used for wine. The inside of the fruit (outside the kernel) is eaten, expressed for oil, used as a medicine or for dressing the hair in various parts of Nigeria.

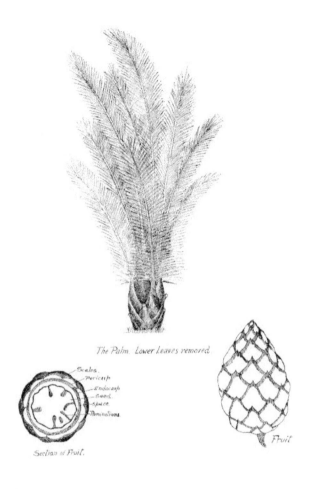

The Palm. Lower Leaves removed

Scales
Pericarp
Endocarp
Seed
Space
Paminations

Section of Fruit.

Fruit

RHUS INSIGNIS Del.—*Kasheshi.* ANACARDIACEAE.

A small tree of the open Bush savannah, some 15-20 feet high. There is either a single stem or several. The crown is distinctive with its long, straight, slender branches and light appearance. It is very susceptible to fire and is frequently met with in coppice form owing to the destruction of the stems. The distinguishing features are the drooping leaves in threes and the small black, wrinkled fruits.

The Bark is grey or light brown, vertically, but not deeply, ridged.

The Leaves are borne in threes at a node and the short stalk bends back, allowing the leaves to hang vertical. They are 3 inches long and an inch broad, a rich, dark green above, silvery below, with the dense covering of silky hairs.

The venation is parallel on either side of the mid-rib, short and long nerves alternating. The leaf is inclined to fold up along the mid-rib.

The Flowers appear in the rains and are in large, erect sprays. Each flower is about ¼ inch in diameter with 5 sepals, 5 white petals and 10 stamens. Owing to their numbers and prominent position at the twig ends, they are conspicuous.

The Fruits are numerous, flattened, single-seeded ovals, shiny black with wrinkled skins and persistent for many months on the leafless tree in the dry season. They are very hard and are a ready means of identification.

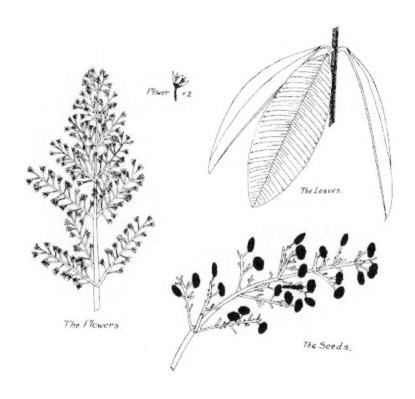

Flower × 2

The Leaves.

The Flowers

The Seeds.

SARCOCEPHALUS RUSSEGERI Kotschy.—*Tafashiya*. RUBIACEAE.

This common species is often more of a shrub than a tree, but in the more favourable localities which it inhabits, chiefly "fadammas" or wet soils among rocks, 30 feet is by no means an unusual height. Older trees generally have several large, crooked stems rising from a common stock or a short, stout bole some 5 feet in girth. Smaller, shrub-like specimens are very straggling with open growth and long, drooping and interlacing branches. The crown, in spite of the size and thickness of the leaves, is fairly open. It is easily identified by either the flowers or fruit which latter can be found for several months of the year.

The Bark is grey, or brown and dull in older trees, with very deep, wide and long fissures and thick, soft ridges. The scales are up to 12 inches in length. The slash is yellow, with crimson streaks.

The Wood is a deep red-brown colour, but is not used owing both to the small sizes and the lack of bole in the tree.

The Leaves are large and rounded with a tongued tip. They average 7 inches long and 3 inches wide, but reach as much as 10 inches long and 5 inches wide. The upper surface is a dark, shining green with pale venation; the under surface pale green with the venation raised. There is a short, red leaf-stalk.

The Flowers are in balls about 2 inches in diameter and appear in February. They fade very quickly. The flower-balls are composed of some hundreds of small flowers which seem white, as each is a yellow, tubular, 5-lobed corolla with a shiny white pistil with an acorn-shaped stigma which is very long. There are 5 small stamens attached to the inside of the corolla, which do not appear. The flowers are sweet-scented.

The Fruits vary somewhat in size and shape but are roughly round or oval, 2-3 inches in diameter and dark red-brown with the surface pitted with the pentagonal scars of the flowers. There is a crimson flesh surrounding a mass of small seeds round a pith centre. They are edible and have a not unpleasant, sweet, acid flavour. They ripen towards the end of the year but can be found for many months on the tree and fall to rot on the ground.

Uses.—A medicine for stomach pains is made from the fruits.

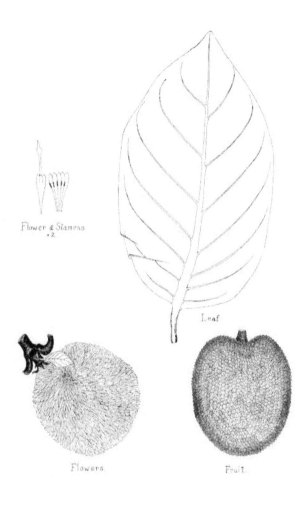

Flower & Stamens
× 2

Leaf

Flowers.

Fruit.

SCLEROCARYA BIRROEA Hochst.—*Danya.* ANACARDIACEAE.

This very common tree was, until 1922, considered a species of *Spondias*, from its resemblance to *S. lutea,* a species occurring commonly further south. It is of medium size, up to 40 feet in height with girths of 6-8 feet. It grows on almost any soil and flourishes in loose sand or dry barren situations where it can be found almost as pure forest. The short heavy bole gives a large volume of timber, but the bole length rarely exceeds 12 feet from which point the much bent branches form a large, round, open crown giving little shade. The pinnate foliage, red flower-spikes or yellow plums distinguish it. Owing to its use for native mortars enormous waste takes place, one mortar being carved from trees which will provide three or four, the rest being allowed to rot or burn. Quite large limbs can be pulled off by hand as they snap very readily. It is not very fast growing and seedlings branch at ground level. Seedlings, however, are very hardy, sending a thickened tap-root several feet into the ground the first year.

The Bark is light grey, sometimes smooth and silvery, and large scales, which turn up at their edges and give the stem a ragged appearance, leave the stem smooth again after falling. The slash is salmon pink to red and spongy and fibrous in composition.

The Wood has a dirty white ground colour with a well-marked reddish grain in bands and streaks, with dark brown patches. In transverse section the pores are small, not very numerous, evenly distributed, mostly single with a few double pores and small nests, in oblique rows, the soft tissue very poorly developed as fine lines imperfectly connecting the pores. The rays are fine, closely and regularly spaced and fairly straight, and they show as light-reflecting bands in radial section, and as a fine stippling in tangential section. The wood is soft, coarse-grained, sawing roughly and picking up, often full of small knots and liable to small borer-beetle attacks. It is fairly durable if well seasoned and weighs 36 lbs. a cubic foot. Good sizes are obtainable.

The Leaves are pinnate, some 9 inches long with 8-9 pairs and a terminal leaflet, rounded, with the venation much branched and prominent on both sides. They are bluish-green with a slight bloom and tend to fold up along the mid-rib.

The Flowers appear from January to April and even in May in the extreme north, on small, 2-3 inch long, erect spikes on the leafless twigs. They are slightly scented. Each has 4 small red sepals, 4 petals recurved, green with red tips, 16 yellow stamens and a red pistil. Each flower is in the axil of a small, red bract.

The Fruits ripen from April to June and are light yellow or pale green plums 1½ inches in diameter, with a tough skin and a juicy, mucilaginous flesh which is with difficulty separated from the rough stone.

Uses.—The wood is greatly in demand for mortars. The fruit is eaten fresh but its amenability is increased by stewing, the flesh being full of fibres attached to the stone.

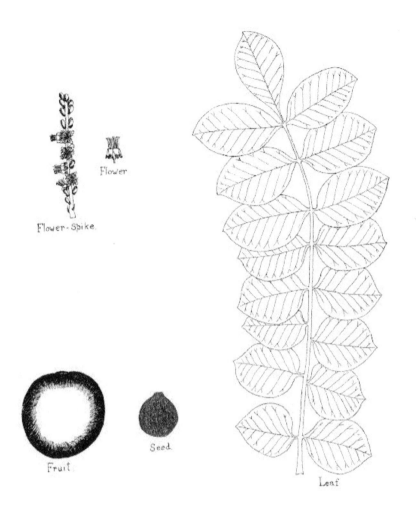

Flower

Flower-Spike.

Seed

Fruit.

Leaf

SECURIDACA LONGIPEDUNCULATA Fres.—*Uwar magunguna, Sainya.* POLYGALACEAE.

This common shrub or small tree is very familiar in open bush and scrub country, growing in the driest localities up to the extreme north of the country. As generally met with it is a small, erect and delicate shrub, some 12-15 feet high, but trees up to 25 or 30 feet are not uncommon, especially in Sokoto province, where a girth of 3 feet is attained. It is readily distinguished by its flowers and by its seeds, and the slender, erect branches are distinctive. It is common amongst the broken rocks of the Bauchi Plateau.

The Bark is fairly smooth and light brown, yellowish or grey, with very small dark-coloured scales at certain times. The slash is yellow.

The Wood is light yellow in colour and the annual rings are very distinctively marked in dark brown, the dry wood parting at these rings into a series of concentric cylinders. Weight 55 lbs. a cubic foot.

The Leaves are small and elliptical in shape, about 2½ inches or less in length, and under ½ inch wide. They are a light grey-green on both surfaces and are set spirally round the long twigs, erect and nearly parallel with the twig.

The Flowers are irregularly shaped, of an uniform red-purple, in small loose spikes. Two of the sepals are modified and appear to be petals. There is a concave lower petal with an appendage and two small petals. There are 8 stamens with peculiar-shaped anthers. The flowers appear from February to April and are very decorative and highly perfumed.

The Fruits are winged seeds, 2 inches long, the seed portion being prominently ridge-veined and the wing with close parallel venation. They become red before drying to pale yellow and are conspicuous.

Uses.—The root, which has a rank odour, is made into a concoction and taken as a purge. Bits of it are also worn as a charm. The seeds are crushed and used as a substitute for soap and a concoction of them taken as a cure for colds. The roots are sometimes an ingredient of arrow poisons. The name "Mother of Medicines" is given it from the number of medicinal properties possessed by the plant.

Seed.

Leaves.

Front

Flowers

Side

STERCULIA TOMENTOSA G. & P.—*Kukuki*. STERCULIACEAE.

A common tree which inhabits granite country and may be found in quantity on hills or in valleys amongst rocks. It does not occur in the extreme north on laterite formations. There is rarely any length of bole, the stem branching low down and the large, crooked limbs spreading wide apart to form a very open crown of irregular form. It can be distinguished at once by its purple bark. Amongst rocks the roots are above ground for several feet, clasping the boulders or creeping between the crevices. Root flanges are common. The branches are so soft in the wood that they, or even the whole of a fair-sized tree may be swayed to and fro by the hand, without difficulty.

The Bark is purple and quite smooth except for occasional large grey scales which leave yellow patches of very distinctive appearance. A gum exudes from the crimson slash and a watery sap flows at certain seasons.

The Wood is white and so soft as to be useless.

The Leaves are about 4-5 inches long and 3½-4 inches broad, cordate with the lobes overlapping the stalk, roughly 3-lobed, the middle lobe being the most prominent and the outline of two other lobes at the base being sometimes visible. There are 5 main veins corresponding with these lobes. The leaves are light green and downy on both sides, the nerves prominent on the under surface. There is a stalk about 3 inches long. The leaves are soft in texture.

The Flowers which spring from old as well as young wood are in small erect panicles 2-3 inches long, appearing February to April. The calyx is ½-¾ inches across, pale green with reddish lines from base to near tip, 5-lobed and downy. There are no petals. The stamens are on a column which is divided into 5 branches, each bearing 3 anthers. The anthers surround the style which is curved and finally the stamens drop off and leave the style exposed.

The Fruits are pods about 3-4 inches long and 2 inches in diameter. There are 4 or 5 together radiating from a twig end and they are the readiest means of identification. They are rounded in the middle and pointed at the end with a groove on the under side along which the pod splits. They are covered with greenish hairs like plush, in a manner similar to the fruits of the Baobab tree. About December the fruit ripens and splits, the seeds falling and the pod remaining on the tree. The seeds, about a dozen, are purplish with a horny, yellow aril at the base and they are attached to both edges of the pod, sitting on small bosses which are covered with short stiff brownish hairs that penetrate the skin of the fingers. There is a large roomy space in the pod, the seeds being small in proportion.

Uses.—The watery sap which exudes from a slash in the spring is drunk in extremity of thirst.

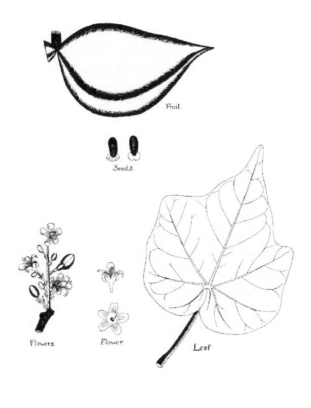

Fruit

Seeds.

Flowers.

Flower.

Leaf

STEREOSPERMUM KUNTHIANUM Cham.—*Sansami, Jiri.*
BIGNONIACEAE.

This is, as commonly met with, quite a small tree, some 20 feet high, with a girth of 2 feet. Larger specimens up to 60 feet in height with a girth of 4 feet are occasionally seen, these dimensions showing the slender proportions of the species. It occurs in clumps, specially fine examples of which are to be seen in Sokoto where they are spread by root suckers. The stem is nearly always waved or spiral, this peculiarity persisting in the larger trees. The stem forks early and the twisted branches form a high crown and the clumps a light canopy under which little grows. Twin stems are common and root suckers abound. The power to grow these suckers is very persistent, as evidenced by cases of them appearing on land which has been farmed for many years and on which there has been no sign of the tree being allowed to grow. Great difficulty has been experienced in growing the tree by artificial means though the seed has been germinated, but the seedlings would not stand transplanting.

The Bark is pale grey or greenish with very large scales which expose contrasting light patches after the manner of the Plane tree. The slash is white with green edges.

The Wood is white or cream-coloured, with tinges of yellow and pink. In transverse section the rings are indistinct darker lines, the pores are very small, in festoons with a large percentage of soft tissue plainly visible as flecks and long concentric lines. The rays are not very closely or evenly spaced, seen as light-reflecting bands in radial section. The grain is fairly straight, following the waved stem of the tree, and coarse with the numerous open pores. The rings show as bands in radial section. The wood is fairly hard, easily sawn, not so readily planed, picking up a little, the finished surface being rather coarse. The weight is 60 lbs. a cubic foot.

The Leaves are pinnate with 2-3 pairs of leaflets and a terminal one. The whole is about a foot long and the leaflets 3-4 inches long. They darken and toughen with age and are greyish-green beneath and much darker above. The lower pair are often rounded at the tip. The mid-rib is almost white in contrast to the upper surface, and the venation is raised on the underneath.

The Flowers are in large, drooping panicles of beautiful, pale pink, funnel-shaped blossoms, which appear in March and bloom till the leaves are grown. They are sometimes much darker on individual trees. They have a small 5-lobed calyx, dark at the base, a pink, tubular, 5-lobed corolla with the lobes crinkled and covered on the inside with hairs and darker lines on the inside of each petal. They have four stamens, grouped on each side of the pistil, 2 longer and 2 shorter, with the appearance of a lyre.

The Fruits are pods about 18 inches long and ¼ inch thick, cylindrical and spirally twisted with a long tapering tip. The pod is filled with a whitish pithy substance into which the seeds with wing above and below are embedded, the wings shaped concavely to fit against the pith. The pods are dark brown and split down both sides throughout their length, the seeds sliding down and falling out. The pods are so persistent that their halves are found on the twigs at the same time as the next year's pods are fully grown.

Uses.—In large sizes in Bornu it is used for mortars, and cut for fuel and "gofas," being most conveniently forked. Locally, pagans will lay a length of it across the house door to prevent thieves entering. There is a superstition that smoke from it will cause leprosy and in Sokoto it is called Dan Sarkin Itache as a mark of respect.

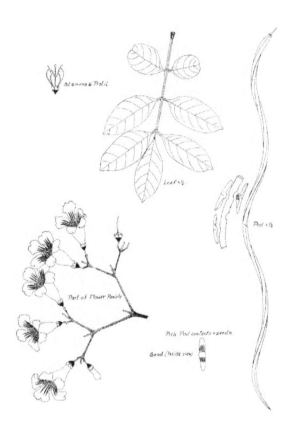

Stamens & Pistil

Leaf ×⅓

Pod ×½

Part of Flower Panicle

Pith Pod contents + seeds

Seed (Inside view)

STRYCHNOS SPINOSA Lam.—*Kokiya*. LOGANIACEAE.

This is the best known and most familiar of the Strychnos species and can be distinguished from the others by the large size of its fruits. It is usually a small tree about 15-20 feet high but will grow over 30 feet in height with a girth of about 3 feet. The stem branches low, the branches ascending at first and then bending over and drooping low, especially when borne down by the weight of the fruits. This drooping habit distinguishes it from the other species which occur in the open. Often the branches interlace after the manner of the *Zizyphus* species. The species occurs everywhere in open situations and on the most barren soils.

The Bark is smooth and light brown in colour, and small, grey scales form and leave light patches when they fall. The slash is yellowish, with green edges.

The Thorns are in pairs at rather wide intervals along the branches and twigs, and are white with black tips. They are sharply recurved.

The Wood is whitish with grey streaks, hard, close grained and sound, picking up a little under the plane, finishing with a smooth surface, and weighs 65 lbs. a cubic foot. In transverse section the pores are very fine, numerous in the many chains and festoons of soft tissue which are clearly visible to the unaided eye and form almost a herring bone pattern. The rays are long, nearly straight, and variable in thickness.

The Leaves are oval and about 2½ inches long and 1½ inches broad with a short stalk. They are shiny on the upper surface, lighter beneath, and have palmate venation of a distinct type.

The Flowers, which appear in February, are in little clusters, and are green. Each is ¼ inch long and has 5 narrow sepals and a bell-shaped corolla of 5 lobes, whose mouth is closed by a ring of hairs. They are slightly perfumed.

The Fruits, which distinguish this species from the others, are large, yellow, round berries, some 4 or 5 inches in diameter, green when unripe, bright yellow and resembling an orange when ripe. The hard rind encloses a number of round flat seeds in a brownish, sweet, edible pulp. The seeds themselves are poisonous. The fruit is very slow in maturing and may be found on the tree for about six months in the year.

Uses.—The pulp of the fruit is eaten, the seeds being rejected as poisonous.

Leaf.

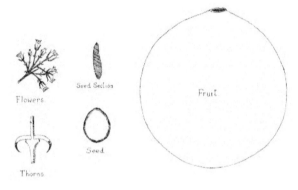

Flowers. Seed Section Fruit.

Thorns. Seed.

STRYCHNOS TRICLISIOIDES Baker.—*Kokiya*. LOGANIACEAE.

A small, erect tree, from 12-20 feet high, or more, with a girth of 1-2 feet. The branches are vertical, and do not spread, there being no crown as such. The stem may be free of branches for some 10 feet, or there may be more than one stem, forking near the base. It does not occur far north, and is particularly common on rocky soils in granite country. It is in form very dissimilar to the other well-known species, *S. spinosa*, and unlike it has no thorns. It can be determined by its flowers, fruits and leaves, which are all typical of the genus.

The Bark is smooth in all shades of pale greys and greens, with a powdery surface. The small scales leave light, concave scars like the Plane tree. The bark is so distinctive that when recognised, it will identify the tree alone. The slash is cream-coloured, with green edges.

The Wood is yellow, hard, coarse and stringy, difficult to plane, cracks when drying and weighs 55 lbs. a cubic foot. In transverse section the rings are indistinct, the pores large, closed and strung rather widely apart between or on the rays which are long, fine and wavy and close together, varying much in thickness.

The Leaves are in pairs, each pair at right angles to the next above and below though a twisting occurs to enable the leaves to turn their faces to the light. The leaf is 2-3 inches long and 1-1¼ inches wide, oval in shape, narrow at the base, broad at the tip, a dull, smooth pale green above, paler beneath, the venation prominent on the under side only and composed of 3 main nerves, the laterals leaving the main rib about a third of the way up. The secondary venation is cellular and delicate. The stalk is dorsally flattened.

The Flowers are typically in threes on short stalks and are borne in small clusters, especially on last season's wood below the new season's leaves, in April-June. Each has a cup-shaped 4-lobed calyx, a tubular, 4-lobed, yellowish corolla, 4 stamens with anthers which protrude slightly from a ring of hairs closing the mouth of the corolla and a pistil whose black stigma also protrudes.

The Fruits are a small edition of those of *S. spinosa*. They are spherical, 1½-2 inches in diameter, ripening from a shiny blue or grey-green to a bright yellow, with a thick "shell" in which are some 6-9 seeds, each in its own sweet, sticky, orange-coloured pulp. The seeds are ½ inch in diameter, rounded, flattened, concave one side, convex the other, not hard, and readily cut in two. The fruit has a small, blunt "nose," and ripens about June.

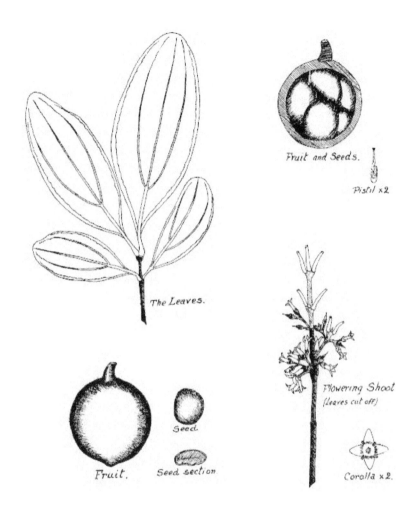

Fruit and Seeds.

Pistil x2

The Leaves.

Seed

Fruit.   Seed section

Flowering Shoot
(leaves cut off)

Corolla x2.

SWARTZIA MADAGASCARIENSIS Desv.—*Gamma fada, Bayama, Bogo zage, Gwazkiya.* LEGUMINOSAE.

This is generally quite a small tree, some 15-20 feet high, but will attain a height of 35-40 feet with girths of 4-5 feet or more. It is erect, with straight stem and ascending branches which in older trees terminate in pendulous twigs from which hang the long, soft, pinnate leaves. Its range does not extend very far north and it is particularly common in Bauchi, both on the plateau at some 4,000 feet altitude and in the lower plains.

The Bark is dark grey and shaggy with long ragged scales which fall in large pieces. That of small trees is light grey and smooth. The slash is yellow.

The Leaves are pinnate, with alternate or opposite leaflets, averaging 11 in number, the whole some 6-8 inches long, the short-stalked leaflets oblong with rounded base and rounded, slightly cleft tip, 2-3 inches long and an inch or more wide. The upper surface is smooth, the lower covered with silky hairs. The colour is rather pale green.

The Flowers are most distinctive with a large rounded, wavy-edged petal, white, with greenish-yellow base and silky hairs on the back. The globular calyx splits into cupped sections as the flower expands, the stamens, varying in number, some 15 or so, are yellow and the ovary and pistil is curved. The whole is on a stalk up to 2 inches in length. The flowers are solitary or in small racemes.

The Fruits are pods from 6-12 inches long, straight, curved or bent, and about ¾ inch thick. They are cylindrical and have a smooth, shining, hard shell, dark brown in colour. The interior is divided transversely into a number of cells in which the brown, flattened seeds lie with their width in the direction of the pods' length. The space on each side of the seeds is full of sticky matter and the pod has an objectionable odour, and persists for a long time on the tree, not splitting.

Uses.—A concoction from the pods is used as a fish poison.

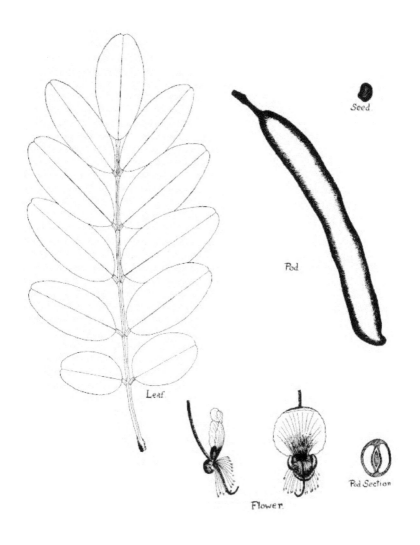

Seed

Pod

Leaf

Flower

Pod Section

TAMARINDUS INDICA Linn.—*Tsamiya. "Tamarind Tree."* LEGUMINOSAE.

One of the commonest species of the north, in park formations, farm lands and open country round towns. It thrives on poor soils, big specimens growing on loose sand and shallow laterite. It is readily identified from a distance by its very dense, dark, compact crown, which may be oval or cylindrical with a pointed top, and uniformly dense down to within a few feet of the ground. It has as a rule a short, stout bole, 6-8 feet in girth, often divided a few feet from the level of the ground into two or three large limbs, with short, gnarled and crooked branches and a dense thicket of twigs. This compact form is more common farther north, southern specimens showing a higher, wider, and more umbrella-shaped crown. It averages 40-50 feet high. It is to be seen at times growing in a curious manner in small clumps of four or five stems on a raised earthy mound, possibly an old ant hill or an accumulation of earth held up by the close-growing stems. There is a very considerable leaf-fall and little grows beneath its shade. A quite common association is that of this species with *Adansonia digitata*, the Tamarind in this case embracing the other with long, sinuous limbs to a considerable height, and having no main stem.

The Bark is light grey with even-sized, thick scales about 1 inch square. These scales extend to the ends of the branches, getting smaller and more regular, still thick and hard, with a considerable resemblance to the bark of the Shea Butter Tree. The youngest trees show this bark. The slash is pale red with a yellow outer layer.

The Wood. The heartwood is dark brown, almost purple-brown, the sapwood pale yellow. In transverse section the pores are large, in groups and festoons, the contents nearly filling them, and they are numerous and the soft tissue occupies a large percentage of the wood. The rays are very fine and close, not visible to the naked eye. The wood is tough, very hard, heavy, cross-grained, blunts axes, saws with difficulty and picks up under the plane, which will give a hard finish, taking a polish. It is liable to crack in seasoning. The weight is 58 lbs. a cubic foot.

The Leaves are pinnate, from 4-6 inches long with 10-12 pairs of dull dark green leaflets, strap-like and notched at the tip. They have a grey bloom and are waxy to the touch.

The Flowers are in slender, drooping panicles and are 1 inch across with 4 yellow sepals, 3 orange-veined petals, 3 stamens and a hairy ovary and pistil. They appear from December onwards.

The Fruits are pods which may be any shape from a single-seeded, round knob to a sickle-shaped, or straight, fleshy pod containing, when dry and ripe 6 or more hard shiny, brown seeds varying in shape from flat and oval to rounded cubes. The ripe pod is yellowish and falls entire. They ripen at the end of the year or later.

Uses.—The wood is used for implement handles and fuel.

The flowers are eaten fresh or as sauce.

The pods are used for making a bitter, not unpalatable drink, which is a laxative.

The young saplings of this species are those used by the Filani for the tests of endurance by flogging undergone by their youths.

In hollows in the trunk are found cocoons of a silkworm, *Anaphe* sp. (tsamiya) which are boiled with water and ashes, dried in the sun and spun into the silk thread used for ornamenting the best gowns.

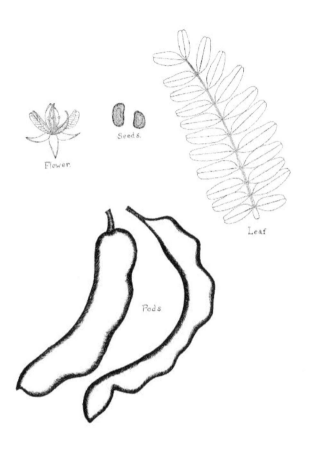

Flower.

Seeds.

Leaf

Pods

TERMINALIA AVICENNIOIDES Guill. & Perr.—*Baushi*.
COMBRETACEAE.

This is one of the commonest trees in a certain type of savannah, of medium quality and height, and it is associated with other species of the same genus, confusingly alike, over hundreds of square miles of forest, especially round about the parallels of 11° and 12° N. The leaf, though it is one of the means of identification, is very variable in width, so that more than one species seems to be included under the single name. With other *Terminalia* species it often forms over 50 per cent. of the forest and occurs pure in small groups. Usually a small tree, 15-20 feet high, it attains over 30 feet, with a short bole and widely spreading branches forming a large and very open crown. The distinguishing feature is the cottony felt covering of both sides of the leaf, the unequal basal lobes, and the downy fruit.

The Bark is grey or blue-grey, with very deep fissures and prominent vertical, wavy ridges of hard cork, typical of the genus. The slash is yellow, rapidly darkening on exposure.

The Wood resembles that of Oak in colour and hardness and often in grain, being light brown with splashes of lighter tone, specially noticeable on the tangential section, with a bright sheen. In transverse section the rings are clear and the pores are clearly visible to the unaided eye as lighter than the ground colour, grouped in concentric rings of soft tissue, or evenly distributed in numerous festoons. The rays are very fine, closely spaced and slightly wavy. The wood is hard, sound, carpenters well and weighs 55 lbs. a cubic foot.

The Leaves are 7-9 inches long and 2-5 inches broad, either elliptic or long and narrow, the upper surface dark green and softly cottony, the under side almost white with the nerves very prominent and the network of veins clearly outlined. The basal lobes are often unequal and the tip more often rounded. The type illustrated is twice as broad as the other extreme found.

The Flowers are in spikes about 4 inches long, loosely grouped, white or pinkish in colour, the pink on the tips of the sepals. The calyx is pubescent, and there are 8 erect stamens. The flowers are scented and appear from February to May.

The Fruits are winged seeds 2-2½ inches long and 1 inch broad, tapering at the ends and covered with a purplish or grey bloom.

Uses.—The roots are used for bows and walking sticks.

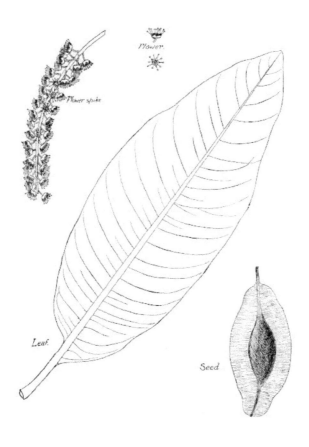

Flower.

Flower spike

Leaf.

Seed

TERMINALIA GLAUCESCENS Pl.—*Baushi*. COMBRETACEAE.

This is a large tree, sometimes as much as 60 feet in height with a good clean bole 6-8 feet in diameter, especially on stream banks. The crown of large trees is high, rounded and open, and a bole length of 30 feet is not uncommon. The distinguishing feature is the smooth under surface of the leaf, which has only a slight pubescence, and the dark smooth surface of the upper side. It is a prominent feature of better class savannah and one of the largest species.

The Bark is dark grey and has very prominent wavy ridges with deep fissures, typical of the genus. The slash is yellow, darkening rapidly.

The Wood is not unlike that of *T. avicennioides,* but lighter in colour and more twisted in the grain, carpentering badly. It is heavier, up to 65 lbs. per cubic foot, and is much coarser.

The Leaves are about 6 inches long and 3 inches broad, oval or elliptic, rounded at the base, variable at the tip. Both surfaces are smooth, the under surface having a slight down. The nerves are prominent beneath and the reticulation of the veins very fine and clear.

The Flowers are on spikes 4-5 inches long, white, with rather long flower stalks, a 5-pointed calyx, pointed in bud, 10 erect stamens and a long ovary. They appear in the leaf axils from February to May and are highly scented.

The Fruits are the typical winged ones of the genus, some 2¾ inches long and an inch wide, thus unusually long for their breadth. They ripen through brilliant shades of red and white to brown, with a fine mauve or grey bloom.

Uses.—The roots are used for bows and walking sticks.

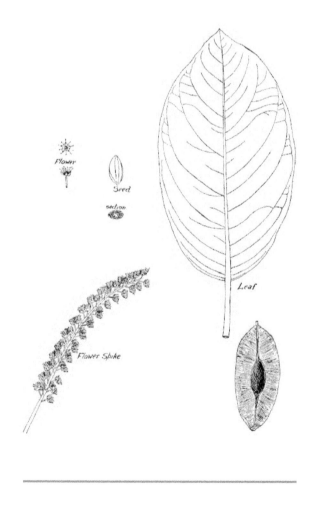

Flower

Seed

section

Flower Spike

Leaf

TERMINALIA MACROPTERA Guill. & Perr.—*Kandari.*
COMBRETACEAE.

One of the largest-leaved species and more readily distinguished on sight than most species. It reaches a height of some 40 or more feet with a girth of 6-7 feet. The bole is short, the heavy branches crooked and spreading wide to form a rounded crown, fairly regular in form. The species forms gregarious clumps and is not so evenly distributed through Terminalia forest as are the other species. In the open savannah of the better type which it inhabits the contrast between the large pale green leaves and the almost black stem is very marked. The distinguishing features are the large leaf, smooth on both surfaces, the very marked rosettes of leaves, the short or almost absent leaf-stalk and the large fruits.

The Bark is dark grey or almost black on the bole. The scales are large, varying from rectangular to square and in old trees form long ridges and fissures in waved pattern. That on the branches is light grey and the long vertical scales are sharply cut transversely. The slash is a dark red brown.

The Wood is similar to that of *T. glaucescens*, light oak-brown with an uneven grain, difficult to carpenter and weighing some 60 lbs. a cubic foot.

The Leaves are a foot or more in length, larger on smaller trees, obovate with tapering base, broad blade, sharp tip, wavy margins and very short thick stalk. They are pale green, almost the same colour on both surfaces, smooth above and below, the heavy mid-rib almost white and prominent on both sides, as are the very pronounced nerves. They grow in marked rosettes which face the light all round the tree and thus form quite good, though superficial, shade. Their texture is tough.

The Flowers are found from February to May, in long spikes in the axils of the new leaves. They are white with the usual 5-pointed calyx, 10 stamens and short pistil. The calyx is smooth but there is a slight pubescence on the inflorescence as a whole.

The Fruits are the usual winged seed, a large and heavy crop of which is borne on the lower part of the spike, the rest seedless. The seed is 4-5 inches long and 1-1½ inches broad, pale green at first and smooth, turning a rich purple with a bloom, the wing slightly spiral twisted, the seed itself tapering acutely at both ends. It dries brown.

Uses.—The roots are used for sticks and bows.

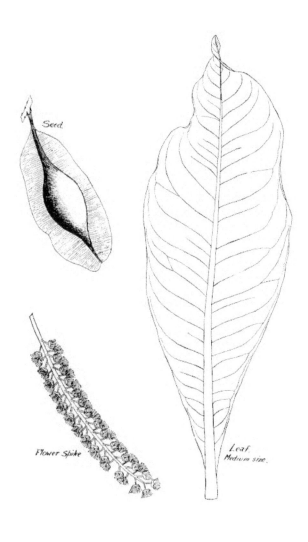

Seed

Flower Spike

Leaf.
Medium size.

TRICHILIA EMETICA Vahl.—*Gwanja kusa, Jan saye.* MELIACEAE.

A medium-sized tree occurring up to 11° N. in Tree or better-class Bush savannah. In Zaria and Bauchi provinces gregarious clumps are not uncommon. A height of 30 feet with a girth of 3-4 feet is an average well-grown tree. The distinguishing features are the large, soft pinnate leaves, green flowers and crimson capsular fruits, the last very conspicuous. It is a low-branched, spreading tree, giving good shade and apparently grows very slowly.

The Bark is pale grey and rough with large, thick, corky, rectangular scales.

The Leaves are in terminal tufts on the thick twigs, at first erect, then drooping, and are 15 inches long with a heavy stalk covered with velvety hairs. The leaflets average 4 pairs with a terminal leaflet and increase in size from below upwards, the basal pair being nearly round and the others increasing in length, with a tapered base and broad cleft tip. The mid-rib of each leaflet projects slightly just below the cleft tip, not actually in the cleft. The upper surface is a rich, smooth green, the underside paler, with prominent nerves, sunk on the upper surface of the leaf.

The Flowers are nearly an inch in diameter, appearing from February onwards. They are in short, stout racemes with a hairy bract at the base of each flower. They appear before the leaves and flower till the leaves are almost full grown. Each has a calyx of 5 sepals which cups the corolla of 5 long greenish petals whose face is concave and tips curved as in the bud. The stamens are joined for half their length and form a ring round the pistil whose globular stigma surmounts them.

The Fruits are prominent capsules, ripening to a rich red colour before drying brown. They are 4-segmented with an uneven surface and a prominent "nose," and contain 4 cells with a seed in each, not always distinctly separated or fully developed. The fruits are most conspicuous amongst the leaves.

Uses.—The local uses are few, the oil being rarely expressed, and the root occasionally used as a mild purge, unpopular owing to its extreme bitterness. The oil is, however, valuable from other parts of Africa, and has been exported, the seeds containing some 60 per cent. of oil suitable for soap and candles and has been valued at £9 10s. per ton of seed.

Section of Fruit.

Fruits.

Leaf. ½

Flowers.

UAPACA GUINEENSIS Müll. Arg.—*Kafafogo*. EUPHORBIACEAE.

A large tree up to 40 feet high with 6-8 feet girth. It occurs as almost pure forest in some localities, is very local, requires good soils and rainfalls, and in such conditions does not, as a rule, exceed 30 feet high with 3-4 feet girths. It is especially common in S. Sokoto, Kontagora, on the escarpments of the Bauchi Plateau and in parts of Zaria Province and on the Plateau itself attains large sizes on the banks of small streams. It is most easily recognised by its shining rosettes of leaves, wax-like yellow flowers or yellow fig-like fruits. It has an erect stem, short bole, whorled branches and regular rounded crown, not very dense from the shade point of view. Its leaf fall covers the ground and little grows beneath it.

The Bark is black, with small, round, even-sized, closely packed scales, often lichenous. Very old trees lose these scales at the base and a brown, fibrous looking bark replaces them. The small black scales are borne almost to the tips of the branches. The slash is red.

The Wood is pale red. In transverse section the rings show as fine white lines between broader, darker bands. The pores are small, numerous and in rows between the long fine rays which are visible as small bands in radial section. In the plank the pores appear numerous and short and the grain is marked faintly in dark and light lines. The pores glisten with resin. The wood is not hard, easy to work and planes to a soft finish with no polish. It is liable to radial cracks but is otherwise a sound, clean timber.

The Leaves are up to a foot long and 6 inches wide, oval, rounded at the tip and gradually tapering to the short, broad stalk. They are a dark, brightly shining green, much paler beneath, covered with short, erect, stiff hairs, feeling slightly rough to the touch. The mid-rib is raised a little on the upper surface and is very prominent beneath. The nerves are light in colour.

The Flowers are monoecious, male and female on the same tree, the female situated below the male. Both are similar in appearance, grouped in clusters round the thick shoots at the base of the leaves, and their parts are surrounded by thick yellow bracts in place of petals. The male has dense clusters of separate flowers, compressed into a ball, each minute flower having 5 2-anthered stamens and a rudimentary ovary amongst minute bracts. The female has 3 triple stigmas. The flowers appear during the rains.

The Fruits are like figs at first sight, but very different in structure. Borne in long clusters on the leafless twigs, or amongst the leaves, they ripen in March or April and may persist till July. They are ¾ inch in diameter, round or slightly pear-shaped, yellow and juicy ripening to dry and red-brown. The remains of the stigma show at the tip as three small black, shredded tufts radiating from the centre, and the division of the fruit shows itself on the outside by three slight ridges. A cross section through the fleshy fruit shows the design of the curiously folded green cotyledons lying in the kernels. The fruit coat is covered with short stiff hairs. The seeds are hard, ½ inch long, oval and grooved.

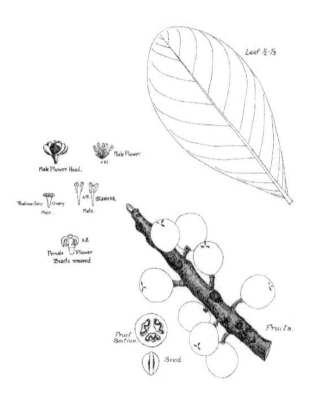

Leaf ½-⅓

Male Flower

Male Flower Head.

Rudimentary Ovary
Male.

Stamens.
Male.

Female Flower
Bracts removed

Fruit Section

Seed.

Fruits.

VERNONIA AMYGDALINA Del.—*Shiwaka.* COMPOSITAE.

A very common shrub or small tree, abundant on the banks of streams and large rivers where it forms an impenetrable thicket. It has either a number of stems from ground level or a single low-branched stem forming a rounded crown up to 12 or 15 feet high.

The Bark is light grey and smooth with no distinctive character.

The Leaves are some 6 inches long and 1¾ inches wide, this being an average, the largest reaching 8 inches in length and 3 inches in width. They are alternate on the shoots, long and tapering at both ends with an acute point and stalks about ½ inch long. The margins are waved. The upper surface is a dull dark green, paler beneath with the venation prominent. There are often hairs on the surface.

The Flowers are in large terminal panicles, sometimes nearly a foot wide. The flowers are in composite heads about ½ inch long, with some 20 flowers in an involucre of pale green scales, the corolla and bifid style white. The flowers are found mostly in the rains.

The Fruit is a seed, or achene, with the usual pappus or ring of bristles by which it is distributed on the wind. The "thistle-down" heads are about ½ inch in diameter.

Uses.—The root is used as a chew-stick for cleaning the teeth and as tonic bitters. The leaves are used medicinally.

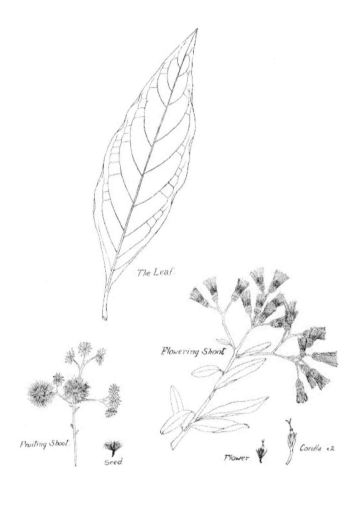

The Leaf.

Flowering Shoot

Fruiting Shoot

Seed

Flower

Corolla ×2

VITEX CIENKOWSKII Kotschy & Peyr.—*Dinya, Dunya.*
VERBENACEAE.

A large tree which occurs more or less evenly distributed throughout the forests. It is rather exacting as a rule as to soil conditions and will not flourish in poor situations, preferring deep soils. It grows to a height of 30-40 feet with girths of 6 or 7 feet, and specimens 10 feet in girth are by no means uncommon. It can be distinguished at a distance by its very large, dense, dark green and regular crown with rounded top. The crown, in the case of trees grown in the open, comes to within 6-10 feet of the ground. Examples growing in the forest exhibit a distinct form, with longer boles and higher, more cylindrical crowns, and the branches ascend at an acute angle to the stem. It is a prolific fruiting species, but more often than not the fruits are burnt on the ground by the grass fires which occur from December on, the months when the fruits fall. It grows very well from the seeds which retain their germinating qualities for some time.

The Bark, which is a very distinctive feature, is light brown or grey, appearing smooth at a distance, and is fibrous in appearance with very narrow and long, vertical fissures and stringy ridges. It is often much darker on the limbs than on the stem. The slash is yellow, darkening on exposure.

The Wood is light brown, rather like Teak. In transverse section the rings are indistinct darker lines, the pores are large, open, mostly single, with a few double pores. The rays are plainly visible, broad, waved, unevenly spaced, with sometimes room for two pores between. The soft tissue round the pores is very poorly developed. In vertical section the pores are distinctive for their curves and the rays are plainly visible as brown flecks in tangential and iridescent bands in radial section. Some of the pore contents glisten. The wood is soft, coarse but nicely grained, very easy to saw and plane, with a soft finish and no polish. The weight is 53 lbs. a cubic foot.

The Leaves are digitate with usually 5, sometimes 6, lobes. They are about 8 inches across, but vary considerably in size and have a 4 inch stalk. A round, pointed leaf-gall is common on the leaves, which are dark green with a bloom on the upper surface and paler with raised venation on the under surface. The young leaves are a deep red-purple.

The Flowers are in small, erect panicles which appear in January. They are about $3/8$ inch across, tubular with 5 petals, 4 of which are small and white, the fifth larger and pink. There are 2 long and 2 short stamens.

The Fruits are in pendulous bunches which ripen from December on. They are black and shiny with little brown lenticels on the skin, a black or dark brown flesh and a large stone. They are $3/4$-1 inch in diameter. The calyx, which enlarges and holds the unripe fruit, dries, shrivels and recedes from the ripe fruit. Bees are greatly attracted by it.

Uses.—The wood is used for making small canoes and large drums.

The fruits are used for making sweetmeats (alewa and madi) and have a taste of honey.

The young leaves are eaten with groundnuts, pepper and salt: they are also an ingredient of ink, being mixed with gum and boiled down to a thickened extract.

Beehives are commonly placed in the branches of this tree.

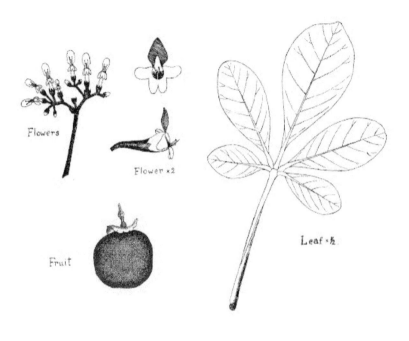

Flowers

Flower x2

Fruit

Leaf x½

VITEX DIVERSIFOLIA Baker.—*Dunyar biri.* VERBENACEAE.

A small tree some 15 feet high, occasionally more, not very common but occurring here and there in Tree savannah and with a very wide distribution. It has little form, being erect and of open growth. Its distinguishing character is the highly scented leaf and the fact that some may be simple and others tri-foliolate. The leaves and branches are in whorls of three, and the fruits like small, oblong editions of those of the well-known *V. Cienkowskii.*

The Bark is light grey and very thick with soft cork and large rectangular scales. This is, however, largely due to fires.

The Leaves are 4-5 inches long and 2-3 inches broad with a stalk 2-3 inches long. When there is a single leaf it is rounded ovate, with slightly cordate base, but if tri-foliolate the leaflets are obovate with tapering base. The margins are waved. The upper surface is dark green with hairs scattered all over the surface and the under side is densely covered, especially on the venation, with hairs, this being far more marked when the leaf is younger, when the venation is very prominent, though it is grooved on the upper surface. The leaves are generally in whorls of three at each stem node. When crushed they give off a strong sage-like odour.

The Flowers are in cymes in the leaf axils, on long stalks, either in whorls of three or pairs. Each has a 5-pointed bell-shaped calyx, an irregular tubular corolla, the lip large and mauve, the remaining lobes greenish and pubescent, 4 stamens pressed up against the top of the corolla and a small, curved white pistil with bifid stigma.

The Fruits are egg-shaped, about ¾-1 inch long, ripening from green to black, spotted with lenticels and cupped for about one-third of their length by the hairy enlarged calyx. They are fleshy drupes with a hard "stone" containing 4 cells, each with one seed. The flesh is very narrow.

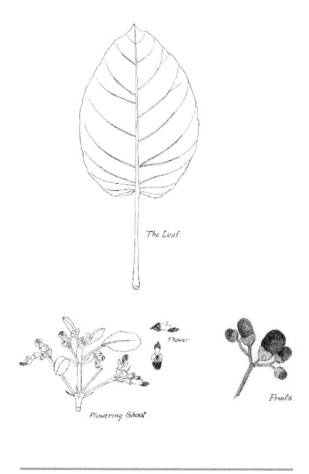

The Leaf.

Flower.

Flowering Shoot.

Fruits.

XIMENIA AMERICANA Linn.—*Tsada*. OLACACEAE.

A small shrubby tree some 10-15 feet high, widely distributed all over the forests and sometimes found in great quantities over small areas. It does not grow in the very dry sandy or rocky northern regions. It has one or several stems from a common stock which spread out and form an open low crown of no form. It can be readily distinguished in fruit and flower.

The Bark is dark brown or almost black, with very small, close-fitting rectangular scales. That of the branches is dark grey and smooth. The slash is crimson and fibrous.

The Thorns are about ½ inch long, slender, very sharp, straight and grey in colour.

The Leaves are 2 inches long and ¾ inch wide, tapering at both ends and with pointed tip: sometimes rounded with slightly cleft tip. The mid-rib has a tendency to curve back, the leaves folding up along it. The surface is smooth and the stalk short and stout and curved.

The Flowers, which appear in January are in small axillary clusters of a dozen or so, each having its own slender stalk growing from the tip of a stouter main stalk. There are 4 small, slender, pointed sepals, 4 petals whose tips are recurved and whose inside surfaces are covered densely with soft, erect hairs, 8-10 stamens inside the corolla, and a pleasant perfume which can be detected from a great distance.

The Fruits are like yellow cherries, with thin skin, sweet flesh and hard stone. The style remains as a prominent pointed tip to the fruit, which is very subject to grubs that may destroy the whole crop.

Uses.—The fruit is eaten and is, with its acid sweet taste, very refreshing.

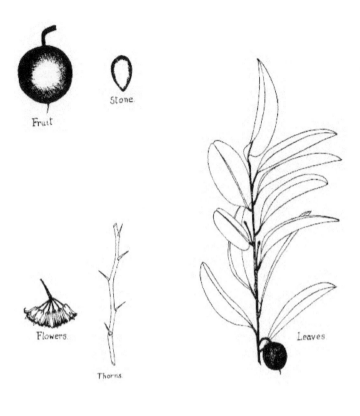

Fruit

Stone

Flowers

Thorns

Leaves

ZIZYPHUS JUJUBA Lam.—*Magariya*. RHAMNACEAE.

This species bears a marked resemblance to *Z. Spina-Christi*, but differs essentially in the under-surface of the leaf having a dense pubescence. It is, as a rule, a smaller tree, though examples may be found almost as large as the other species, 40 feet high with a girth of 5-6 feet. The crown is similar, with a dense tangle of armed twigs, which interlace and droop to a considerable distance. The twigs are very slender and zig-zag sharply at the nodes. This species, unlike the other, is not often seen in towns, but occurs locally in plenty in the forest, in the more open and drier localities especially. It is, like the other, the favourite haunt of small birds, which build their nests in association with those of a small wasp. The crown is not so dense as that of *Z. Spina-Christi*, since the leaves are smaller and further apart.

The Bark is a smoky grey colour with a brown tinge and has narrow fissures and long stringy ridges and scales. The slash is cerise pink.

The Thorns are in pairs, the upper ¾ inch long and straight, the lower shorter and sharply recurved. They are a pale orange colour, and they and the light grey twigs are covered with a fine pubescence. The upper, straight thorn differs from that of *Z. Spina-Christi* in being set at near right-angles with the twig, not sharply inclined forward.

The Wood is pale red, in transverse section darker. The sapwood is white. In transverse section the rings show clearly as white bands in the red ground, the pores are small and single and the rays extremely fine, invisible to the naked eye. In vertical section the pores are fine, the grain indistinct. The wood is fairly hard, saws well and planes readily without picking up, to a smooth finish which will polish. A good sound wood with occasional knots.

The Leaves are about 1½ inches long and ¾ inch broad, are alternate and assume one plane with the upper surface outwards. They are dark, shiny green above, with whitish venation, and thickly covered with a grey pubescence beneath. The veins are palmate, three spreading from the base, raised beneath. The margins are very finely serrate.

The Flowers are in small axillary bunches and are downy, greenish, with a flat, 5-pointed calyx, 5 minute petals, and 5 minute stamens and a bifid stigma. The clusters are not so numerous or large as those of the other species. They appear about January.

The Fruits are small, reddish drupes about ½ inch in diameter, with a crisp, whitish flesh, which is sweet and edible, and a stone which is very large for the size of the fruit. The fruit is smaller and drier, and the stone rounder and larger in comparison with it, than those of the other species, *Z. Spina-Christi*. It ripens about December.

Uses.—The fruits are eaten fresh.

Fruit

Leaves.

Seed

Flower ×2

Flowers

ZIZYPHUS MUCRONATA Willd.—*Magariyar kura.* RHAMNACEAE.

This species is a small, irregular-shaped tree not, as a rule, above 20 feet in height with a girth of 2-3 feet. It is quite commonly met with as a shrub-like plant, some 10-12 feet high with straggling branches forming a tangle. There may be two or three stems springing at ground level from a common stock. The crown is open and uneven, and the long drooping branches extend to within a few feet of the ground.

The Bark is grey, with long, coarse fissures and rough scales.

The Thorns are in pairs, the dorsal one long and straight, the lower shorter and recurved. They are brown in colour.

The Leaves are dark, dull green, soft on the upper surface, grey and densely covered with velvety hairs beneath. They are about 2 inches long and 1½ inches wide, with short stalks and have finely serrate edges, which the other species have not. The venation is 3-palmate and raised on the under surface. They are borne alternately on the twigs, with a tendency to assume one plane, the upper surface outwards.

The Flowers are in small axillary clusters in the axil of the leaf between the thorns, and are greenish-yellow with a 5-pointed calyx, 5 minute petals, 5 stamens, a bifid stigma and a yellow receptacle. They appear in December.

The Fruits are similar, but larger than those of the other species and are a rich red-brown, with a shiny, brittle skin, a white, mealy flesh and a large stone. They are not edible.

Uses.—The fruits are chewed, but not swallowed, as a cure for toothache. The wood is used for making bows of superior quality.

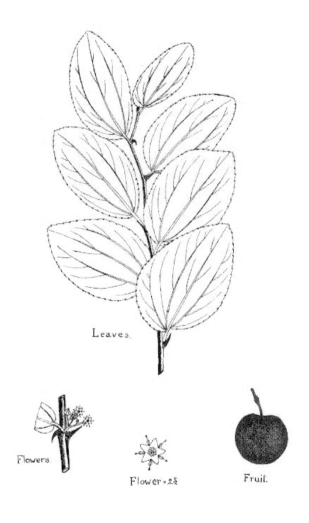

Leaves.

Flowers.

Flower × 2½

Fruit.

ZIZYPHUS SPINA-CHRISTI Willd.—*Kurna*. RHAMNACEAE.

This species is generally found in and around towns where it attains a height of over 40 feet with girths of over 6 feet. Closely resembling *Z. Jujuba*, it can be distinguished at once from that species by the absence of white on the underside of the leaf. The bright green foliage and tangled crown, commonly spherical in young trees, are readily recognised. Large trees have a great thicket of long, slender twigs that intertwine and emerge here and there from the crown to a distance of several feet, after the manner of the Bramble. It is a good shade tree and this and its edible fruits are the reasons for it being planted and preserved in towns. The tree is a favourite nesting site for small birds which build their nests in association with a wasp whose "honeycomb" guards the entrance to the nest.

The Bark is grey and deeply scoured with long fissures and ragged, ridged scales which fall in large sections 6 inches long. The slash is cerise.

The Thorns are in pairs on the white twigs. The upper thorn is long and straight and the lower is shorter and sharply recurved. They are light brown and the leaf rises between them with the flowers in the leaf axil. The straight thorn points well forward.

The Wood is pale red, often with wide, brown discolorations; the sapwood is white. In transverse section the rings are only faintly marked; the pores are very small, numerous and single: the rays, invisible to the unaided eye, are extremely fine and close together, showing as small reddish bands in radial section. In vertical section the pores are long and fine, the grain close, and the colour banded in faint shades of the ground colour. The wood saws and planes well, is sound, though liable to knots, fairly heavy and moderately hard. The weight is 50 lbs. a cubic foot.

The Leaves, though they are arranged spirally round the twigs, tend to assume one plane. The upper surface is bright green with a bluish bloom and the venation roughens the underside. They are 1-2 inches long and ¾-1 inch wide, the palmate venation composed of three principal nerves, spreading from the base.

The Flowers are in small clusters in the leaf axils, on a common stalk which divides into a number of more slender stalks each bearing one flower. Each is ¼ inch in diameter, with a flat, 5-pointed calyx, 5 minute petals and 5 stamens, and a bifid stigma round which is a raised disc which turns brown. The colour of the flower parts is green, and the flowering season is from October to January.

The Fruits are small, light brown drupes about ¾ inch in diameter with a sweet, edible flesh, juicy, then turning dry and mealy, and a hard stone of large size for the fruit. There is often a very heavy crop and they resemble cherries, ripening from November onwards.

Uses.—On account of its strength and supposed immunity from white ants the wood is used for the rafters of flat-roofed mud houses.

The fruits are eaten fresh. The leaves, besides providing fodder for goats and cattle, are crushed and applied to cure a skin disease.

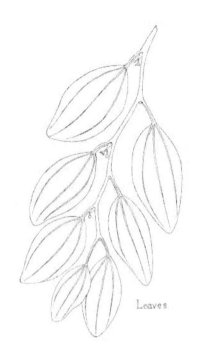

Leaves.

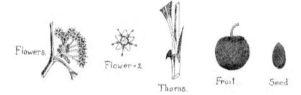

Flowers. Flower×2. Thorns. Fruit. Seed.

APPENDIX I.
NOTE ON THE TABLE OF FLOWERING SEASONS.

The following list of Flowering Seasons has been compiled from 3 years' observations between 9° and 14° North Latitudes. No account has been taken of out-of-season flowering, which is of casual occurrence and due to various causes. The lengths of the flowering seasons are due chiefly to the wide distribution of the trees and do not necessarily represent the lengths of time individuals will bear flowers. Few species, in fact, bear flowers for longer than one month; many for only a few days, but a journey from north to south will show the effects of varying humidity on the flowering season. Such an influence may be counteracted by the date on which a fire swept through the country, fires exerting a quickening influence on the flowering to a marked degree.

The arrangement of the monthly columns so as to place the New Year in the middle of the page is made for the purpose of grouping the flowering period of those species which start flowering before January, whose periods would otherwise be broken up. Comparatively so few trees flower after June that the breaking of their periods is of little account.

This list will illustrate, also, the effect of fires at various times of the dry season on the different species, which may be more exactly determined by making an allowance for the situation of a forest in terms of degrees latitude. In this connection attention may be drawn to the large number of species whose fruits or seeds persist on the branches until the rainy season of the following year, and which will be destroyed by such fires. Only those species whose fruit forms and falls in the rains of the year in which it flowered will stand a fair chance of germinating.

Any observed extension, before or after the periods given here can be entered by extending the lines to correspond and suit the particular case.

TABLE OF FLOWERING SEASONS.

No.	Name of Tree.	Jul	Aug	Sep	Oct	Nov	Dec	Jan	Feb	Mar	Apr	May	Jun
1	Acacia albida												
2	Acacia arabica												
3	Acacia campylacantha												
4	Acacia Dalzielii												

No.	Name of Tree.	Jul	Aug	Sep	Oct	Nov	Dec	Jan	Feb	Mar	Apr	May	Jun
5	Acacia nilotica												
6	Acacia Senegal												
7	Acacia Seyal												
8	Acacia Sieberiana												
9	Adansonia digitata												
10	Adina microcephala												
11	Afrormosia laxiflora												
12	Afzelia africana												
13	Albizzia Brownei												
14	Albizzia Chevalieri												
15	Amblygonocarpus Schweinfurthii												
16	Andira inermis												
17	Anogeissus leiocarpus												
18	Anona senegalensis												
19	Balanites aegyptiaca												
20	Balsamodendron africanum												
21	Bauhinia reticulata												
22	Bauhinia rufescens												
23	Berlinia auriculata												
24	Bombax buonopozense												
25	Borassus flabellifer												
26	Boswellia Dalzielii												
27	Bridelia ferruginea												

No.	Name of Tree.	Jul	Aug	Sep	Oct	Nov	Dec	Jan	Feb	Mar	Apr	May	Jun
28	Bridelia scleroneura												
29	Burkea africana												
30	Butyrospermum Parkii												
31	Cassia Arereh												
32	Cassia goratensis												
33	Cassia Sieberiana												
34	Celtis integrifolia												
35	Combretum abbreviatum												
36	Combretum hypopilinum												
37	Combretum lecananthum												
38	Combretum leonense												
39	Combretum micranthum												
40	Combretum verticillatum												
41	Cordia abyssinica												
42	Crataeva Adansonii												
43	Crossopteryx Kotschyana												
44	Croton amabilis												
45	Cussonia nigerica												
46	Detarium senegalense												

No.	Name of Tree.	Jul	Aug	Sep	Oct	Nov	Dec	Jan	Feb	Mar	Apr	May	Jun
47	Dichrostachys nutans												
48	Diospyros mespiliformis												
49	Ekebergia senegalensis												
50	Entada sudanica												
51	Eriodendron orientale												
52	Erythrina senegalensis												
53	Eugenia guineensis												
54	Ficus capensis												
55	Ficus gnaphalocarpa												
56	Ficus iteophylla												
57	Ficus Kawuri												
58	Ficus platyphylla												
59	Ficus polita												
60	Ficus Thonningii												
61	Ficus Vallis-Choudae												
62	Gardenia erubescens												
63	Gardenia ternifolia												
64	Grewia mollis												
65	Guiera senegalensis												
66	Gymnosporia senegalensis												
67	Hannoa undulata												
68	Hymenocardia acida												

No.	Name of Tree.	Jul	Aug	Sep	Oct	Nov	Dec	Jan	Feb	Mar	Apr	May	Jun
69	Hyphaene Thebaica												
70	Isoberlinia Dalzielii												
71	Isoberlinia doka												
72	Khaya senegalensis												
73	Kigelia aethiopica var. bornuensis												
74	Lonchocarpus griffonianus												
75	Lonchocarpus laxiflorus												
76	Lophira alata												
77	Maerua angolensis												
78	Maerua crassifolia												
79	Mimosa asperata												
80	Mitragyne africana												
81	Monotes Kerstingii												
82	Moringa pterygosperma												
83	Ochna Hillii												
84	Odina acida												
85	Odina Barteri												
86	Ormocarpum bibracteatum												
87	Ostryoderris Chevalieri												
88	Paradaniellia Oliveri												
89	Parinarium curatellaefolium												

No.	Name of Tree.	Jul	Aug	Sep	Oct	Nov	Dec	Jan	Feb	Mar	Apr	May	Jun
90	Parinarium macrophyllum												
91	Parkia filicoidea												
92	Parkinsonia aculeata												
93	Prosopis oblonga												
94	Pseudocedrela Kotschyi												
95	Psorospermum senegalense												
96	Pterocarpus erinaceus												
97	Randia nilotica												
98	Raphia vinifera												
99	Rhus insignis												
100	Sarcocephalus Russegeri												
101	Sclerocarya Birroea												
102	Securidaca longipedunculata												
103	Sterculia tomentosa												
104	Stereospermum Kunthianum												
105	Strychnos spinosa												
106	Strychnos triclisioides												
107	Swartzia madagascariensis												
108	Tamarindus indica												

No.	Name of Tree.	Jul	Aug	Sep	Oct	Nov	Dec	Jan	Feb	Mar	Apr	May	Jun
109	Terminalia avicennioides												
110	Terminalia glaucescens												
111	Terminalia macroptera												
112	Trichilia emetica												
113	Uapaca guineensis												
114	Vernonia amygdalina												
115	Vitex Cienkowskii												
116	Vitex diversifolia												
117	Ximenia americana												
118	Zizyphus Jujuba												
119	Zizyphus mucronata												
120	Zizyphus Spina-Christi												
		12	9	7	18	24	39	51	78	85	67	40	26

APPENDIX II.
CLASSIFICATION UNDER FAMILIES.

Anonaceae.—Anona senegalensis.

Capparidaceae.—Maerua angolensis, M. crassifolia, Crataeva Adansonii.

Polygalaceae.—Securidaca longipedunculata, Psorospermum senegalense.

Dipterocarpaceae.—Lophira alata, Monotes Kerstingii.

Malvaceae.—Adansonia digitata, Bombax buonopozense, Eriodendron orientale.

Sterculiaceae.—Sterculia tomentosa.

Tiliaceae.—Grewia mollis.

Simarubaceae.—Balanites aegyptiaca, Hannoa undulata.

Ochnaceae.—Ochna Hillii.

Burseraceae.—Balsamodendron africanum, Boswellia Dalzielii.

Meliaceae.—Khaya senegalensis, Pseudocedrela Kotschyi, Ekebergia senegalensis, Trichilia emetica.

Olacaceae.—Ximenia americana.

Celastraceae.—Gymnosporia senegalensis.

Rhamnaceae.—Zizyphus Jujuba, Z. mucronata, Z. Spina-Christi.

Anacardiaceae.—Odina acida, O. Barteri, Sclerocarya birroea, Rhus insignis.

Moringaceae.—Moringa pterygosperma.

Leguminosae.—Erythrina senegalensis, Ormocarpum bibracteatum, Pterocarpus erinaceus, Lonchocarpus laxiflorus, L. griffonianus, Ostryoderris Chevalieri, Andira inermis, Afrormosia laxiflora, Swartzia madagascariensis, Parkinsonia aculeata, Cassia Sieberiana, C. Arereh, C. goratensis, Bauhinia rufescens, B. reticulata, Berlinia auriculata, Isoberlinia doka, I. Dalzielii, Paradaniellia Oliveri, Afzelia africana, Tamarindus indica, Detarium senegalense, Burkea africana, Parkia filicoidea, Entada sudanica, Prosopis oblonga, Amblygonocarpus Schweinfurthii, Dichrostachys nutans, Mimosa asperata, Acacia albida, A. Senegal, A. Sieberiana, A. arabica, A. nilotica, A. Seyal, A. campylacantha, A. Dalzielii, Albizzia Chevalieri, A. Brownei.

Rosaceae.—Parinarium curatellaefolium, P. macrophyllum.

Combretaceae.—Guiera senegalensis, Anogeissus leiocarpus, Combretum leonense, C. verticillatum, C. lecananthum, C. abbreviatum, C. hypopilinum, C. micranthum, Terminalia glaucescens, T. macroptera, T. avicennioides.

Myrtaceae.—Eugenia guineensis.

Araliaceae.—Cussonia nigerica.

Rubiaceae.—Sarcocephalus Russegeri, Adina microcephala, Mitragyne africana, Crossopteryx Kotschyana, Randia nilotica, Gardenia erubescens, Gardenia ternifolia.

Compositae.—Vernonia amygdalina.

Sapotaceae.—Butyrospermum Parkii.

Ebenaceae.—Diospyros mespiliformis.

Loganiaceae.—Strychnos triclisioides, S. spinosa.

Boraginaceae.—Cordia abyssinica.

Bignoniaceae.—Stereospermum Kunthianum, Kigelia aethiopica var. bornuensis.

Verbenaceae.—Vitex diversifolia, V. Cienkowskii.

Euphorbiaceae.—Bridelia scleroneura, B. ferruginea, Uapaca guineensis, Hymenocardia acida, Croton amabilis.

Ulmaceae.—Celtis integrifolia.

Moraceae.—Ficus capensis, F. gnaphalocarpa, F. Kawuri, F. Thonningii, F. platyphylla, F. polita, F. Vallis-Choudae, F. iteophylla.

Palmaceae.—Raphia vinifera, Borassus flabellifer, Hyphaene Thebaica.